CONTENTS

THE GUINNESS BOOK OF MILITARY BLUNDERS

GEOFFREY REGAN

yes !
good Health
Happiness for 2000 !
Must be a great year !
Chris
Pederson
x.

Get mo

Please r
You can
Or by f

For Mary Lodge

Published in Great Britain by
Guinness Publishing Ltd
338 Euston Road, London NW1 3BD

"Guinness" is a registered trademark of
Guinness Publishing Ltd.

First published 1991
Reprinted 1992 (three times), 1993 (four times), 1994 (twice), 1995 (twice), 1996, 1997 (three times), 1998, 1999

ISBN 0–85112–961–7 (Guinness Publishing)
ISBN 0–85112–077–6 (The Past Times)

A catalogue record for this book is available from the
British Library

Designed by Cathy Shilling

Typeset by Ace Filmsetting Ltd, Frome, Somerset

Printed and bound in Great Britain by
The Bath Press, Bath

Front cover illustration:
The Battle of the Modder River, 11 December 1899.
Defeat of the Gordon Highlanders under General Methuen
by Boer forces led by General J. H. de la Rey and General
Cronje. (Archiv für Kunst und Geschichte, Berlin.)

INTRODUCTION

The tabloid press has been strangely quiet about the surgeon who fell asleep during the most delicate stage of a heart bypass operation. The local news has given only cursory attention to the bus driver who stopped to take a bath in public, while still responsible for transporting a busload of passengers to their places of work. Even the professional journals have overlooked the architect who thought he was a bird, and the accountant who was unable to rise from his bed in the morning because he thought his legs were made of glass and would shatter. And the case of the solicitor who believed he had been made pregnant with an elephant by a passing soldier was treated with incredulity bordering on ridicule. Why have these cases of misconduct not come to the attention of the public, who deserve to be protected from the disordered excesses of the professional mind? Simply because not one of them is true. All of the offences of which the surgeon, the bus driver, the architect, accountant and solicitor stand accused were in fact committed by soldiers. And not ordinary soldiers, or even extraordinary soldiers, or even nameless individuals whose careers were cut short and whose papers were shuffled and filed and forgotten. No, all of the above cases involved military commanders and generals, men who had reached the highest ranks of the military profession. They are just a few cases, albeit bizarre and newsworthy ones, of the kind of aberrant behaviour which has contributed to military incompetence.

Now it is not my intention to ridicule a profession as vital to the security of the state as the military. That would be as unjust as it would be unproductive. Incompetence is no more a part of a military career than of any other. There are just as many incompetent doctors, dentists, accountants, solicitors, teachers, and engineers as there are military commanders. What makes the failures of the soldiers more significant is the effect that they can have on society. Where an airline pilot can kill hundreds of people, a general can kill tens of thousands. Where a surgeon can amputate the wrong leg, a bomber pilot can spray napalm on his own men rather than the enemy. A legal error, however serious, is rarely life-threatening. An accounting error can cost thousands, possibly even millions of pounds, but short of causing people to leap from high buildings, its effects can be rectified. The damage from a misguided bomb, missile, battle, campaign or war can never be repaired. Military mistakes have a way of proving permanent, and while we can laugh at some of the errors depicted in this book, we must never forget that they were paid for in blood and human suffering.

No general ever sets out to be a failure. He would not have risen to high rank unless he had played by the rules, kept his boots, belt and buckle clean, saluted the right people at the right time and shown that he was good at following orders. Such a man, having done all of these things, might then be considered fit to give orders to others. As writers far more skilled than myself have shown, the capacity to give orders is not dependent on the ability to follow them. Giving orders involves thinking for yourself, not a quality encouraged among soldiers in the lower ranks. The only test of whether a commander can command is to give him the chance, by which time it is too late to do anything about it if he proves to be deficient. A hit-and-miss method of officer selection was used by President Lincoln during the American Civil War. He only settled on Ulysses S. Grant

Once described as the best colonel in the British Army but perhaps the worst general, Sir Redvers Buller is the classic example of the gifted soldier promoted beyond his capacity. Despite his undoubted courage, his tactical shortcomings brought humiliations on himself and his country in the Second Boer War.

as commander after he had proved to his satisfaction that John Pope, Ambrose Burnside and Joseph Hooker were hopeless. The dead of the battles of Second Manassas, Fredericksburg and Chancellorsville might have applauded his final decision if they had lived to see it.

Military promotion sometimes involves a law of diminishing returns, by which able junior officers become less able as they are promoted beyond their level of competence. Examples of this – the 'Peter Principle' – are legion in the military profession. Perhaps the best is that of Sir Redvers Buller, the darling of the late Victorian soldier, who won the Victoria Cross during the Zulu Wars and was an inspiring regimental officer. Yet Buller, a man of action rather than a thinker, suffered the ignominy of seeing his career fall apart during the Second Boer War, ending as a laughing-stock nicknamed 'Sir Reverse Buller'. This tragic process is not confined to the British army alone. The Austrian general Ludwig Benedek was praised for his brilliant achievements during the Franco-Austrian War of 1859, yet by 1866 he came to be universally reviled as a blunderer and a traitor for his failings in the disastrous war against Prussia. The French marshal Achille Bazaine – hero of the Crimean War, of Solferino, and Mexico – was branded a traitor in 1870 as a result of his failures in the Franco-Prussian War. Court-martialled and sentenced to death, Bazaine escaped to Spain and died in poverty there, a pitiful figure promoted beyond his capacity – a victim rather than a villain. If readers find amusement in some of these stories, I hope that they will remember the salutary lessons of Achille Bazaine and Ludwig Benedek, both ruined by their own previous successes – more sinned-against than sinning.

Villains there have been and butchers too. But often these men have been shielded from the horrors of front-line fighting by distance or telephone lines or concrete bunkers. Sometimes the blunderers have been the planners, who have sent men into battle ill-equipped, badly clothed or unfed. They have sent them to fight in deserts, in mountains made impassable by snow, in swamps, or in disease-ridden jungles. They have asked men, and their generals, to do things that were impossible, and when the casualty lists have been collected they have blamed the commanders and the frontline soldiers for failings that were rightly theirs.

And seated in leather chairs, behind oak desks and in panelled offices, sit the politicians, the deputies of the people, in whose name the slaughter is committed. Only visiting quiet areas of the front, distanced from the consequences of their policies by the filtering presence of staff officers and commanders desperate to impress, it is easy for the politician to forget that he deals in death just as surely as the man who gives the order to attack. His mistakes are every bit as costly as the general's and his motives frequently far more cynical. If the military Saul has slain his thousands, the political David has certainly slain his ten thousands, and frequently with far less cause.

'Brains! I don't believe in brains. You haven't any, I know, Sir.'

The Duke of Cambridge complimenting one of his generals.

The aged and incompetent Duke of Cambridge was commander-in-chief of the British army from the Crimean War until 1895. He did whatever he could to resist army reform and was particularly hostile to officers who studied their profession. As a result in 1900 50 per cent of all military literature was written in Germany, 25 per cent in France and just one per cent in Britain.

THE BATTLE OF TANGA (1914)

From the outset the British regarded the campaign against the Germans in East Africa in 1914 as a minor operation. It could, they felt, be safely left to their Indian Army. Regrettably the Secretary of State for India was to take this attitude of insouciance a stage further: in his opinion such a trifling venture could be quite happily assigned to second-rate troops. In choosing a commander, however, the British slipped badly from the second-rate standards they had set themselves, appointing a bungler – Major-General Aitken – whom few would have rated so highly. Aitken was a soldier more suited in style and appearance to the colonial campaigns of the nineteenth century. He had a supreme confidence in his own ability and that of his troops. Thirty-five years in India had convinced him that Indian soldiers would soon make mincemeat of a 'lot of Niggers'. He preferred to base his campaigns on prejudice rather than reasoned argument, stressing the weaknesses of 'Blacks' and 'Huns', and refusing advice from anyone professing local knowledge or intelligence.

In fact, no one in the expeditionary force sent from India knew anything of their destination, which was the port of Tanga. During their brief stay at Mombasa, Aitken was offered the help of Lieutenant-Colonel B. R. Graham of the King's African Rifles, but refused, preferring to stick to his Indian troops. When Graham warned Aitken that the German native troops (Askaris) should not be underestimated, the General disagreed, saying they were untrained and that he would thrash them all well before Christmas.

Aitken's description of his troops as 'magnificent' was touching if hardly accurate. Of the 8,000 men under his command, only the North Lancashire Regiment and the Gurkhas were anything other than mediocre. The Indian troops were some of the worst in the Indian Army, being untrained, ill-equipped and poorly led. Some had only recently been issued with modern Lee-Enfield rifles and did not understand how to use them properly. There were soldiers from all parts of India, speaking twelve different languages, following different faiths and commanded by men who, in some cases, had never even seen their units before the embarkation at Bombay. Aitken's Intelligence Officer, Captain Meinertzhagen – despite his name an Englishman – described the Indian troops as, 'the worst in India . . . I tremble to think what may happen if we meet with serious opposition. The senior officers are nearer to fossils than active energetic leaders.'

The soldiers may have been of poor quality, but the treatment they received during the voyage from Bombay to Mombasa served to reduce morale to rock bottom. Because of a delay in sailing, the soldiers spent sixteen unnecessary days aboard the transports in crowded conditions and appalling heat. When they finally embarked no consideration was given to the differences in caste, religion or dietary needs. Most of them spent the voyage either seasick in their bunks, or suffering from diarrhoea brought on by eating food to which they were unaccustomed. When it was suggested to Aitken at Mombasa that he should allow his men ashore to recuperate from the effects of the voyage he pooh-poohed the idea and said that it might alert the Germans. It was safer to take his men straight on to Tanga.

Although the expedition was supposed to be secret, the Germans had every possible warning that it was coming. The labels on the crates in Bombay dockyards announced, 'Indian Expeditionary Force "B", Mombasa, East Africa', and headlines in the British and East African press heralded the imminent arrival of the force. In addition, there were the plain radio messages between the convoy and Mombasa and the letters from German residents in British East Africa to their friends in Tanga. The German commander, Colonel von Lettow-Vorbeck, had a lot to thank the British for. The fleet even travelled down the African coast in sight of land, just in case any Germans might have missed it.

Sailing ahead of the main force, the cruiser HMS *Fox* under Captain F. W. Cauldfield entered Tanga harbour to discuss the status of the town with the German Governor, von Schnee. Von Schnee had been in the habit of arranging truces with the Royal Navy in order to save Tanga from bombardment. But with Aitken's convoy not far behind, Cauldfield had come to tell the Germans that all truces were cancelled. Von Schnee was missing, but the local commissioner, Herr Auracher, told the captain that he would need time to consult higher authorities. The ingenuous Cauldfield then asked Auracher if the harbour was mined and, not surprisingly, the German replied that it was full of mines. Leaving the trusting naval officer waiting, Auracher rushed off to send a message to Colonel von Lettow-Vorbeck that the British had arrived. He then donned an army uniform, raised the German flag, and went off to join his military unit. After a while, Cauldfield began to suspect that Auracher was not coming back so, returning to the *Fox*, he ordered a tug to begin the laborious task of minesweeping. When the rest of the convoy arrived they had to endure the frustration of waiting while a variety of logs, oil cans and, for all we know, old boots, were swept from the harbour. There were, in fact, no mines, but Auracher had won valuable time for von Lettow-Vorbeck to begin entraining his troops for Tanga.

The landing and its aftermath could easily have come from the pen of Evelyn Waugh (it was recently fictionalized by William Boyd in *An Ice-Cream War*). Cauldfield, convinced that there were unknown hazards at Tanga, persuaded Aitken to land at a point a mile farther down the coast, out of sight of the town. This proved, in fact, the worst possible place to land, being a mangrove swamp full of leeches and water snakes and covered by a miasma of mosquitoes and tsetse flies. Into this horror the miserable Indian troops were plunged, 'fresh' from their experiences on the voyage. Little wonder that they were ready to jump at every shadow and panic at every sound. Meinertzhagen, landing with the first troops at 10 pm, made himself a bed in the garden of a dwelling designated 'The White House', with a mattress filled with 'nice bits of lingerie' taken from the house, and blankets consisting of a large Union Jack and a German flag.

By the time the British troops were fully ashore the Germans had had 48 hours in which to make their preparations. As soon as Aitken ordered the advance on Tanga a number of things started to go wrong. Although outnumbered by eight to one, von Lettow-Vorbeck was not without hope, remarking on 'the clumsiness with which English troops were moved and led in battle'. As the British advanced towards Tanga through the cocoa plantations they could not see any Germans waiting for them. In frustration, three British officers climbed up a small hill to see better and were immediately shot dead. Suddenly, a German bugle was heard and the Askaris rushed to attack the 13th Rajputs, who simply turned and ran, leaving their twelve British officers to be killed on the spot. When Meinertzhagen tried to stop the panic, an Indian officer drew a sword on him and had to be shot. Brigadier Tighe, commanding the Bangalore Brigade, signalled to Aitken, watching from the deck of one of the ships, that his men were facing 2,500 German rifles. In fact there were just 250 Askaris. This first assault had cost the British 300 casualties, mostly officers and NCOs. So panicky were the Indian troops by this stage, that when a rifle went off by accident 100 Rajputs rushed all the way back to the beach, some of them standing up to their necks in the sea.

Meanwhile, the British transports had been landing masses of military supplies on the beaches, irrespective of whether they were needed or not. The scene was one of pandemonium. Since no scouting was taking place none of the British had any idea of the Germans' position or numbers. Aitken had in any case decided to use his full strength in the next attack. Spearheaded by his best troops, the North Lancs and the Gurkhas, with the Indian regiments bringing up the rear, he renewed the attack on Tanga. Curiously enough, he had refused the offer of a naval bombardment to soften up the German positions, partly because he did not want to damage civilian property, but mostly because he did not actually know where the Germans were and did not like to admit it.

The Germans had set up a strong defensive position, linked by field telephones and fronted by barbed wire. There were snipers in the baobab trees and machine guns at intervals on the ground. It was a formidable challenge for even the best troops. But by this time many of the Indian soldiers were in a state of collapse from heat stroke or thirst, having already drunk the contents of their water bottles even though it was only midday. As they approached their invisible enemies the Askaris shouted insults such as 'Indians are insects'. The Imperial Service Brigade, which contained the weakest units, found themselves wading through fields of corn eight feet high, while unseen Askari snipers in the trees drilled holes in the tops of their heads. The Indians were also terrified by the clouds of smoke issuing from the black-powder rifles of the Askaris.

In one part of the field matters were going better for the British. The North Lancs and Gurkhas had routed the Askaris and captured the customs house and hospital in Tanga, marking the fact by raising a large Union Jack. But elsewhere the Indians were about to face an even greater enemy than the Germans. Hanging from the trees across the battlefield were hives of hollow logs containing African bees, a particularly large and aggressive form of the insect. Infuriated by the noise and the bullets, the bees emerged in clouds from their hives and descended on the advancing Indian troops. At once panic spread, with troops fleeing towards the safety of the sea pursued by bees, which stung them as they went. Refreshingly impartial, some bees stayed to sting the Askaris, but the main attack was directed at the British troops. One engineer was stung 300 times while another, unconscious from a wound, returned to consciousness to find himself being dive-bombed by hundreds of bees. To the hysterical British soldiers it seemed as if the bees were yet another cunning German trick. Even *The Times* later wrote that the bee hives had been used as weapons by von Lettow-Vorbeck. When asked about this the German merely smiled, saying 'Gott mitt uns.'

On board the HQ boat the appearance of hundreds of British troops on the beach, waving their hands above their heads and leaping into the sea, must have been an astonishing sight. 'You don't suppose they're being driven back?' asked a bright staff officer. On the beach one British officer could hardly bear to report the cowardly behaviour of his troops, remarking, 'I would never have believed that grown-up men of any race could have been reduced to such shamelessness.'

Furious, Aitken ordered an immediate naval bombardment, which had to be stopped when the only hit recorded in Tanga fell on the hospital, crammed with British dead and wounded. The other shells fell into the retreating British lines, causing further casualties. The Indian troops were shooting so wildly that they were doing more damage to their own side than to the Germans. One North Lancs soldier commented, 'We don't mind the German fire, but with most of our officers and NCOs down and a bloody crowd of niggers firing into our backs and bees stinging our backsides, things are a bit 'ard.'

The British suffered over a thousand casualties in this one attack and overall lost 800 dead, 500 wounded and 250 missing. Against this von Lettow-Vorbeck's losses were light – 15 Europeans and 54 Askaris killed and wounded. His success was complete, the more so when Aitken promptly re-embarked his troops, abandoning all their equipment. After the British had gone von Lettow-Vorbeck was able to equip new regiments with British rifles and machine guns and had enough food, coats, blankets, motorbikes, telegraph equipment and other supplies to last him for a year.

The evacuation of the wounded was arranged by Captain Meinertzhagen, who found the Germans magnanimous in victory. 'You English', they told him, 'are really quite incomprehensible. You regard war as a game.' As if to prove the point a group of men from the North Lancs Regiment amused themselves by swimming in the sea while the evacuation was taking place, horrifying the Germans with this breach of protocol. What the Germans would have thought of the British sailors who rowed into Tanga harbour at the height of the fighting, hoping to buy some food in town, is anyone's guess.

But the travails of General Aitken and his Indian troops were not ended. Arriving somewhat chastened back at Mombasa, they were refused permission to land by customs officials unless they paid a five per cent *ad valorem* tax. It was the bayonets of the North Lancs Regiment which managed to convince the customs men that 'Expeditionary Force "B" had come to stay. But for General Aitken there was no happy ending. Kitchener, the Secretary for War, refused to see him on his recall to Britain and he was reduced in rank to Colonel and retired on half pay.

CHAPTER 1: UNFIT TO LEAD

The Aged

Generals, unlike fine wines, do not necessarily get better with age. The military profession can be an extremely arduous one and as high rank is usually only achieved after middle age, it is quite possible that the ageing process will have exacerbated defects of both mind and body. It is now generally accepted that hardening of the arteries of the brain can lead to loss of memory, confused thinking, emotional instability, general irritability and senility. Any of these weaknesses, natural as they may be in elderly gentlemen of any profession, can have the most disastrous consequences in a military commander.

Tilting at windmills

Probably the most difficult septuagenarian who held military command – and the competition has been stiff – was the Spanish general Gregorio Garcia de la Cuesta. During the Peninsular War, the Duke of Wellington described Cuesta as 'obstinate as any gentleman at the head of an army should be'. A British officer called him 'a perverse, stupid old blockhead'. Wellington would certainly have agreed with this comment about Cuesta's behaviour at the battle of Talavera in 1809, but the interests of Anglo-Spanish co-operation prevented him from saying so.

The main problems were that Cuesta had an aristocratic disdain for foreigners, considered himself a master of the military art and regarded Wellington as a mere beginner. As the two Allied commanders discussed tactics before the battle of Talavera, Wellington found an interpreter redundant, since he soon learned the Spanish for 'no'. The French army under Marshal Victor was approaching the River Alberche and Wellington understood from Cuesta that the Allied forces would fight him there. By two o'clock in the morning the British troops were ready, but there was absolutely no sign of Cuesta and his Spaniards. Hours passed, the French army withdrew, and Wellington eventually rode over to the Spanish camp, where he found Cuesta asleep on some cushions, and full of nonsensical excuses when he was woken. Cuesta insisted he would fight the next day. Wellington summed up his behaviour as 'whimsical perversity'. He clearly thought Cuesta too old and too incompetent to command a modern army.

But once Cuesta decided to act everything changed. He became as reckless as he had previously been cautious. The following day, ignoring Wellington completely, Cuesta's army set off towards Madrid, in pursuit of Victor's army, now reinforced and double the size of the previous day. To the amazed British observers it seemed that Cuesta's army was like 'the last army of the Middle Ages pouring out to do battle with the French Revolution'. There could only be one outcome and Wellington knew it. When the Spaniards caught up with the French they found themselves hopelessly outnumbered and so Cuesta turned his army round and fled back towards the River Tagus, where Wellington was waiting for him.

The British had established defensive positions three miles behind the river, but Cuesta simply refused to cross. He was in a kind of 'no man's land', with the river at his rear and facing certain disaster if the French attacked. Wellington actually went down on his knees to beg Cuesta to take up position behind the river, and eventually he agreed. Even so, Wellington knew that the Spanish army could not be relied on except to occupy space. When it came to the fighting the British would be on their own.

On 27 July, the combined French armies of King Joseph Bonaparte and Marshals Victor and Jourdan faced the Anglo-Spanish at Talavera. Even now Cuesta's men could not help turning the battle into a farce. When French skirmishers fired some casual shots at the Spaniards the result was astounding. The whole front line of the Spanish army fired a volley and suddenly two thousand of the Spaniards, frightened by the noise and smoke of their own guns, fled with a shout of 'Treason', knocking over Cuesta's coach, in which the old man was forced to travel due to his decrepit state. They then rushed into the British camp to begin looting. Thousands of Spaniards fled down the roads into Portugal, spreading the news that the French had won. At the sound of the huge Spanish volley, Wellington, imperturbable as always, turned to his liaison officer saying, 'If they will but fire as well tomorrow the day is our own; but as there seems nobody to fire at just now, I wish you would stop it.'

The armistice that wasn't

The fate of the elderly Austrian general Count Auersperg during the Austerlitz campaign of 1805 shows the danger of leaving matters of vital strategic importance in the hands of an amiable but senile commander. The French desperately needed to cross the Danube at Spitz and knew they must take the Tabor bridge – a long, wooden structure – intact. The retreating Austrians had prepared it for destruction, setting up a mass of guns at its far end, so that the French troops could not try to rush it. The French commanders, Marshals Murat and Lannes, decided to attempt to bluff the Austrians. They walked confidently across the bridge in full view of the Austrian engineers preparing to demolish it, and brushed aside the few Austrian guards. They informed a flustered officer that there had been an armistice and that the bridge was now French. The officer rushed off to see his commanding officer. Meanwhile, French troops were gradually occupying the bridge. Murat walked straight up to the Austrian guns and proceeded to sit on one of them. The scene was now set for the appearance of Count Auersperg, who talked to the French marshals for a while, eventually agreeing to remove his troops and guns, and let the French have the bridge. Auersperg's blunder contributed greatly to the eventual French triumph at Austerlitz, and the hapless old man was court-martialled and sentenced to be shot. Reprieved by the Emperor, he spent his last days in prison.

The Church Militant

It has often been said that old soldiers never die, but just fade away. Unfortunately for the men under their command too many old soldiers do not know when to retire. One such case was Brigadier Pope, who, by the time of the battle of Chillianwalla in 1849, fought in India between the British and the Sikhs of the Punjab, was in his dotage, nearly blind and so feeble he had to be lifted into his saddle. The British military historian Sir John Fortescue describes the action of placing fine cavalry regiments in his care as like letting a small child play with valuable porcelain. Even the army commander, General Gough, thought him far too old to command the cavalry. On the battlefield Pope soon showed that he did not know what he was doing, facing his men in the wrong direction and giving confused orders. At the sight of some Sikh cavalry Pope announced 'Threes about!' instead of 'Threes right!' causing the British horsemen to wheel about and charge off the field. As they fled panic set in and guns and wagons were overturned. But God was with the British that day in the shape of Chaplain Whiting. Hearing troopers shouting that the day was lost, Whiting managed to halt them and exclaimed, 'No, sir! The Almighty God would never will it that a Christian army should be cut up by a pagan host. Halt, Sir, or as I am a minister of the word of God, I'll shoot you.' He managed to rally the cavalry and the battle was won. Whiting was mentioned in despatches, with a recommendation from Gough that he be made a bishop !

Raglan at the Alma

In the Crimean War it appeared that the British had selected their commanders by recruiting from an old people's home. At 67, the expedition commander Lord Raglan was an amiable old man subject to absent-minded fits. Remembering his youthful service as secretary to Wellington in the Napoleonic Wars he called his Russian enemies the

Lord Raglan, commander of the British forces in the Crimean War, was an amiable and much-loved veteran of Waterloo. Too gentle and absent-minded for the rigours of the campaign, he failed to impose himself on his subordinates.

French – who were now in fact his allies. In one particularly absent moment at the battle of Alma in 1854 he achieved something probably unique in the annals of military history, when he single-handedly captured the centre of the Russian position. Having ordered four British divisions to attack the Russians he trotted off across the river on the right of the British line, passed through lines of French skirmishers, who stood open-mouthed at the curious sight of an old gentleman followed by aides in cocked hats riding blithely through the middle of the battle. He continued up the slope until he reached the top of Telegraph Hill, where he paused to admire the view. Thinking it a good place to site some guns he ordered two cannon to be brought up, with which his staff officers blew up a Russian ammunition wagon. Convinced that they had been outflanked, the Russians immediately withdrew their batteries of guns.

Inactive service

Even judged by the venerable standards of the profession, certain Russian commanders who saw service in World War I must be counted especially infirm. General Wenzel von Plehve, commander of the Russian Fifth Army, was not only senile but actually dying when the war began. His performance was to say the least undistinguished. But age was not always a bar to performance, as in the case of 68-year-old General Sukhomlinov, who, despite his inadequacies as War Minister, married a 23-year-old divorcée. With war in the air, and having recommended mobilization, he then announced he was taking a holiday on the French Riviera. Sukhomlinov was no great military thinker: he asserted that warfare had not changed since 1877 and boasted that he had not read a military manual for 25 years. The nearest he came to expounding a military theory was when he quoted a maxim of the eighteenth-century Russian field marshal Suvorov, 'The bullet is a fool; the bayonet a hero.'

Old generals seem to have been the bane of the army of Frederick the Great of Prussia. One general was so senile that when Frederick asked him to report on a young officer, he mistakenly wrote that he was 'very cowardly' instead of 'very capable'. General von Ramin was so long in the tooth and so unintelligent that he rightly belonged, according to his fellow officers, to 'the age of the Vandals and Huns'. Clearly intelligence or the lack of it was not a significant factor in the promotion of some Prussian generals. The story is told of a Prussian general who went to buy a map and when asked whether he wanted a local map or a general map, replied, 'What a question to ask. I'm a general so I must have a general map.'

Geriatric cavalry commanders have always had to cope with the physical demands of mounting and then riding a horse. In August 1914, the cavalry commander of Rennenkampf's First Russian army, the elderly Khan of Nakhichevan, had to be left behind during the invasion of East Prussia. He was later found crying in his tent because he was suffering so badly from piles that he could not mount his horse. Octogenarians, like the Prussian General von Pennavaire, are rare but not unknown in the military profession. A lifetime of experience had not persuaded this general that he was in the wrong business. Known as 'the anvil' because he was beaten so often, von Pennavaire was a loyal if inept servant of Frederick the Great. At the battle of Kolin in 1757 his incompetent handling of the Prussian cavalry caused it to be swept off the field by the Austrians, turning Frederick's defeat into a major disaster.

During the 1745 Anglo-Austrian campaign in the Netherlands, the Austrian commander was the 75-year-old Marshal Königsegg. Königsegg had pleaded with the Empress Maria Theresa to be relieved of his command so that he could return to the baths of asses' milk he took to relieve his rheumatism, but she refused.

Perhaps we should leave the final word to the astringent wit of General Max Hoffman, who masterminded the German success at Tannenburg in 1914 (although the glory went to the aged titan, General Hindenburg, who became a legendary figure for the Germans). Some years after the war Hoffman showed some friends over the HQ from which the battle of Tannenburg was conducted, pointing out, 'Here the Field Marshal [Hindenburg] has slept, before the battle of Tannenburg, after the battle of Tannenburg and, between you and me, during the battle of Tannenburg also.'

The Ill and Incapacitated

It is pointless to pretend that important historical decisions have not been affected by the physical condition of the person making them. And because of the severe stress and danger of a battle this is particularly true of military commanders. In a sense, no general is ever free of such mental and physical pressures, so it is only worth investigating situations where this factor had a significant effect on the outcome of a campaign.

Stonewall

Perhaps the most remarkable example is that of the Confederate general Thomas 'Stonewall' Jackson, a man in whom Robert E. Lee placed the greatest trust. There were, however, many things Lee did not know about Jackson. In the first place, he was a self-willed, self-righteous bigot, and regarded fighting on a Sunday as a sin. Secondly, he was sometimes so eccentric in his behaviour that certain of his colleagues thought him mad. As a junior officer, on one occasion he wore his greatcoat through a long, sweltering summer because 'he had received no orders to change it'. At the Virginia Military Institute he paced up and down outside the Superintendent's office in a pelting hailstorm because he refused to deliver his report one minute earlier than the appointed time. Such behaviour indicates an obsessive personality. Thirdly, Jackson had a physically delicate constitution, needed more sleep than most people, and was slow to recover from prolonged strain. Before the Seven Days' Battle in 1862, Jackson had carried out his brilliant 'Valley Campaign', manoeuvring his forces up and down the Shenandoah Valley for six weeks, disrupting Union communications and tying down a force of Northern troops many times greater than his own. But the effort had left him exhausted and he was on the point of collapse from 'stress fatigue'. He was in no condition to go straight into another campaign and should have told Lee that this was the case. Perhaps he was unaware of his own limitations and thus said nothing. Lee made his plans on the basis of Jackson playing his normal vital role. But the decisive help Lee was expecting from Jackson never materialized, Jackson having collapsed from a 'depletion of the adrenal cortex'.

On 26 June, Lee had arranged for Jackson's Corps to strike the Union forces at Mechanicsville just after dawn on the second day of the Seven Days' Battle, but by midday he had not turned up. In frustration the impulsive Confederate General A. P. Hill attacked instead, although his forces were heavily outnumbered. If Jackson had joined Hill at this crucial moment the army of the Union commander General McClellan faced collapse. But Jackson did not come. While the fighting raged just three miles away his men, who could clearly hear the gunfire, sat on the ground and smoked. Jackson stood praying on a nearby hill, refusing to speak to anyone during the whole afternoon.

On 29 June, at Savage's Station, Jackson failed to bring his troops into action as planned, spending the day rebuilding a bridge instead of crossing the James River to attack the Union rearguard. The next day, at White Oak Swamp, Jackson kept 25,000 Confederate troops inactive while he slept. The long-suffering A. P. Hill had been left in the lurch by Jackson for the third time in a week and had suffered unnecessary casualties. Jackson's behaviour has been called 'disastrous and unredeemable'. In some armies he would have been court-martialled and shot.

Abdel Hakim Amer

One commander – Egyptian Field Marshal Abdel Hakim Amer – was so affected by the pressures of command that he tried to run into a lavatory to shoot himself. During the Arab-Israeli War of 1967 the Egyptian air force was destroyed on the ground in the first few hours of the conflict and the static Egyptian defences in Sinai were overrun by Israeli armoured forces. Amer kept this information from President Nasser and instead told him that his troops had staged a successful counter-attack. This persuaded the Egyptian leader to refuse a cease-fire offer from Israel and to encourage King Hussein of Jordan to enter the war – disastrous errors which cost Egypt the whole Sinai peninsula, and Jordan the West Bank and Jerusalem. Eventually Nasser was warned by General Mortagi that Amer had 'cracked'. He rushed to Army HQ to find that the story was true and Amer was concerning himself with the siting of a single artillery battery. The reactions of commanders who crack under pressure show a remarkable similarity: they revert to what they feel most comfortable doing, usually operating as though they were junior officers again, free of the responsibilities of overall command. Nasser found Amer in a lamentable state, 'his eyes bloodshot, and swallowing aspirins one after the other', and quite incapable of directing Egypt's armed forces. Amer threatened to commit suicide when it was discovered that he had been misinforming his own government. If Amer's mental collapse – and possible drug dependence – helps to explain his behaviour, it is harder to justify that of the many Egyptian commanders who simply abandoned their units in Sinai and fled back to Egypt. When it became certain that he would face trial, Amer finally committed suicide.

French complaints

The pitiful performance of General Maurice Gamelin as Commander of the French Army in 1940 was probably the result of the treatment he was having for syphilis. During the 1930s Gamelin underwent treatment for advanced syphilis of the central nervous system. This results in personality disorders, dullness of intellect and poor concentration, memory and judgement. Gamelin's memoirs contain evidence of wild swings in mood between elation and depression, with delusions of grandeur and paranoia. Gamelin certainly tried to blame others for his own mistakes, notably his rivals Pétain and Weygand. If France was unlucky in having the unstable Gamelin in command, she was positively cursed in having General Georges as his deputy and virtual commander of the Battle of France in May 1940. Georges was mentally and physically unfit to command such a vast battlefield. Once France's finest soldier, he had not fully recovered from the injuries he received in Marseilles in 1934, when the King of Yugoslavia was assassinated by Croatian extremists. Even in 1940 the effects were still obvious. Faced with the collapse of French resistance, Georges simply 'burst into tears'.

'The experiment has been conclusive. Our method has been tried out. I can assure you that victory is certain. The enemy will learn this to his cost.'

General Robert Nivelle, December 1916

Nivelle's planned offensive on the Chemin des Dames was known to the Germans in advance and resulted in a disastrous defeat, costing France 120,000 casualties in five days. Nivelle was sacked. Widespread mutinies rocked the French Army.

In 1917, the chief of staff to General Nivelle, commander of the French Army, was a Major d'Alenson. Although a brilliant officer, d'Alenson was dying of consumption, and was obsessively determined to achieve something great for his country before he died. Alenson's obsession struck a chord in Nivelle and the two men planned a great spring offensive which they believed would win the war for France. D'Alenson wrote that, 'Victory must be won before I die and I have but a short time to live.' He was willing to sacrifice himself for France and he did not care how many careers he ruined in the process, nor how many French soldiers he sent to their deaths because of his maniacal fervour. Influenced by d'Alenson, Nivelle lost his sense of balance and launched the offensive on the Chemin des Dames ridge, even though the Germans had been thoroughly warned that it was coming. The outcome was 120,000 French casualties and mutiny in the French Army.

Napoleon

No military commander has come under greater medical scrutiny than Napoleon Bonaparte, yet it is doubtful if any of his various disorders had any significant effect on his ability to command before 1807. However, after this date, the cumulative effect of so much campaigning, hasty eating and rough conditions brought about a decline in his efficiency. His temper got worse, probably influenced by colic and later a peptic ulcer. His speed of thought got slower at the very time that his armies were getting larger and more difficult for one man to command. His poor display at Borodino in 1812 may have been caused by a bladder complaint which made riding difficult. At Dresden in 1813 he was thoroughly exhausted and suffering from a severe stomach condition. In the same year at Leipzig he was 'doubled up' by stomach pains and his mind was lethargic. When the pain subsided he simply had to sleep. This need for rest had occurred at Jena in 1806,

By 1809 Napoleon's tactical performance was beginning to show the debilitating influence of complaints such as piles and colic. At the Battle of Wagram he sacrificed Macdonald's Corps in an attempt to break the Austrian lines at a cost of eighty per cent casualties.

when soldiers had to form a square around his sleeping figure during the fighting. After the defeat at Leipzig, Napoleon was quite unable to organize an effective retreat as his need for sleep was overwhelming. Napoleon was suffering from pituitary dysplasia, resulting from an over-active pituitary gland in his youth and its progressive failure as he got older. Of Napoleon's many disorders, his prolapsed piles, which made it painful for him to mount his horse at Waterloo in 1815, are probably the best known. Prolonged campaigning became too much for him and he left too much responsibility to Marshal Ney. The Napoleon of 1815 inspired far less fear in his opponents than he did in 1805 or 1800. Perhaps they had got better, but Napoleon had certainly got worse. At his best he was still incomparable, but his good days were becoming too rare. Blunders, brought on by physical and mental decline, were so common by 1815 that at Waterloo Wellington was able to wait for them to happen.

Faulty vision

Poor eyesight has contributed to more military blunders than perhaps any other physical deficiency. Two eighteenth-century examples illustrate why. In 1757 the Prussian army of Prince Augustus William was forced to turn away from the best retreat route when the road was thought to be blocked by batteries of Austrian artillery. The Austrian guns proved to be a herd of cattle. The same army panicked later on seeing further batteries of guns and burned all their transport and pontoon bridges before it was discovered that the guns were in fact tree trunks. 1757 was a bad year for Prussian eyesight, Field Marshal Seydlitz having trapped a French cavalry force in a hollow near Rossbach only for one of his officers to fail to close the trap as he mistook young fir trees for French infantry advancing to the rescue.

The Insane and the Irrational

The mental health of military commanders has always been difficult to assess. Only in the event of disaster does it occur to anyone to ask if a general was in his right mind in giving a particular order. The pressure of military command imposes greater stress on an individual than almost any other activity. Psychological weaknesses that might have passed unnoticed in civilian life emerge with potentially catastrophic consequences in wartime.

Sir William Erskine

One has to feel a certain sympathy for the Duke of Wellington in having Sir William Erskine as one of his senior commanders during the Peninsular War. On hearing of his appointment, Wellington urgently wrote to the Military Secretary in London, saying that he had always understood that Erskine was insane. The Secretary replied, 'No doubt he is a little mad at times, but in his lucid intervals he is an uncommonly clever fellow; and I trust that he will have no fit during the campaign, though I must say he looked a little mad as he embarked.' This was hardly reassuring to Wellington, particularly as Erskine had twice been confined to a lunatic asylum. In addition, his eyesight was so bad that before any battle he had to ask someone to point him in the general direction of the enemy. At the battle of Sabugal, in 1811, in the absence of other generals, Erskine commanded both the cavalry and the light infantry. His actions, in sending his troops in the wrong direction, saved the French from an even heavier defeat.

Erskine's greatest 'achievement' was at the siege of Almeida, where his actions completely thwarted the Iron Duke's plans and caused even a man as mild as Wellington to explode, 'This was the most disgraceful military event that has yet occurred to us.' The besieged French garrison was allowed to escape because Erskine failed to guard the bridge of Barba de Puerca. Apparently Erskine was dining with a comrade at four o'clock in the afternoon when Wellington's order to guard the bridge reached him. Requested to send some cavalry and a force of infantry, his initial response was to send a corporal and four men. Pakenham, a fellow diner, broke in, 'Sir William, you might as well attempt to block up the bridge with a pinch of this snuff as to place such a party for such an object.' Thinking better of it, Erskine decided to send a whole regiment, wrote out the order for its colonel and then absent-mindedly put the order in his pocket and forgot about it. At midnight, Erskine, feeling through his pockets, found the order, and passed it to Colonel Bevan, leaving this unfortunate officer to take the blame for arriving at the bridge too late to halt the retreating French.

Erskine eventually committed suicide by jumping out of a window in Lisbon in 1813. Found dying on the ground, he asked bystanders, 'Why on earth did I do that?'

Prussian field marshal Blücher's timely intervention at Waterloo completed Wellington's victory. But behind the dashing and successful exterior Blücher was prey to bizarre psychoses, at one time believing he was pregnant with an elephant.

Flights of fancy

Some generals have entertained such strange fancies that their sanity has been doubted, even though in lucid moments their ability to command has not. Confederate general Richard S. Ewell, a bald-headed man with a beaked nose and a habit of cocking his head to one side, occasionally believed he was a bird, pecking at his food and emitting strange chirping noises. His diet of wheat boiled in milk was necessitated by an ulcer, but caused his men to harbour doubts as to his mental state. The famous Prussian field marshal Leberecht von Blücher suffered from the belief that he was pregnant with an elephant, fathered on him by a French soldier. Blücher, who was subject to fits of senile melancholia, also claimed the French had heated the floor of his room so that he could only bear to stagger around it on tiptoe. The luckless Wellington, escaping the attentions of Erskine for a moment, reported that Blücher often told him of his fears. With Blücher incapacitated by his mental problems, the ensuing squabbles between the other Prussian generals, notably Gneisenau and Yorck, contributed to Allied defeats by Napoleon in the campaign of 1814.

Unlike Blücher, who for the most part was an inspiring leader, the Greek General Hajianestis had no such redeeming features. A courtier rather than a soldier, Hajianestis was appointed to command the Greek army in its war with Turkey in 1921. Like Lord Cardigan before him, Hajianestis preferred to conduct his campaigns from the comfort of his yacht, anchored in the harbour at Smyrna, while occasionally enjoying the culinary delights of the waterside restaurants. However, the main problem with Hajianestis was that he was certifiably insane.

His profound mental sickness showed itself in many different ways. For some of the time he claimed that he was unable to get out of bed because his legs were made of glass or sugar and were so brittle that they would shatter. At other times his aides would find him lying quite still assuming himself to be dead. When he was 'active' his orders were confused and contradictory, with the result that Greek morale fell to rock bottom. His dismissal was only a matter of time, yet even this was conducted in a farcical manner. His replacement, General Tricoupis, was only informed of his promotion to commander-in-chief in a newspaper article shown to him by his Turkish captors.

The American Dream

Paranoia, with its violent swings in mood, can be a far more dangerous problem for commanders than mere eccentricity or even insanity. The behaviour of one of America's most famous generals, Douglas MacArthur, during the Korean War provides a striking example of this problem. Paranoia had been an ever-present feature of MacArthur's personality from his early days. He had suffered heavy blows to his self-esteem during his marriage to his first wife, who had cast doubts on his sexual capacities. She had told her brother that MacArthur might be a general, 'but he's a buck private in the boudoir', and at parties waved her cocked little finger at him, telling people that 'Douglas doesn't know what his penis is for except to pee with.' This underlying anxiety was undoubtedly a driving force in this ambitious, forceful, but dangerously unbalanced commander.

When MacArthur was appointed to be UN commander in Korea he was at 70 already too old for the shock of modern war. His moods swung violently from depression to

General Douglas MacArthur was commander of the UN forces in Korea 1950–1. His unpredictable handling of the War led to his dismissal by President Truman.

exaltation, rendering it impossible for his political bosses in Washington to gain a balanced view of the military situation. He achieved brilliant success with the Inchon landing through incredible risk-taking, yet at other times exaggerated the problems facing him and made them an excuse for inaction. He also underestimated his enemy, claiming on his appointment that he could handle the Communists with one hand behind his back. After his success at Inchon his prestige was so great that his military decisions were rarely questioned. It was said that he was 'like a Greek hero of old [marching] to an unkind and inexorable fate'.

MacArthur was determined to free North Korea of Communism, even though this was not the aim of UN intervention. To this end he pushed his forces ever northward, and, in spite of Chinese warnings, approached the Yalu River – the border with China. The Chinese reacted by crossing the river and inflicting a devastating defeat on MacArthur's troops. Between 4 and 9 November, he underwent extraordinary swings of mood, first dismissing the possibility that China would dare to oppose the United States, then panicking that his troops were about to be overwhelmed by Communist hordes. By the end of the five days he was calm again, convinced that he could handle anything that they threw at him. All the while Chinese troops had been infiltrating across the Yalu and were surrounding his positions. Defeat brought on a severe paranoiac reaction, in which MacArthur blamed everyone else, notably the politicians in Washington, and uttered apocalyptic threats that defeat in Asia made the fall of Europe to the Communists inevitable. He demanded the chance to use the atomic bomb against China or to saturate areas of Korea with radioactive waste to deny it to the enemy. Nothing less than a full war with China would do for MacArthur. He acted out the famous saying that 'he whom the gods wish to destroy they first make mad'. Clearly President Truman and the American chiefs of staff could no longer leave so severely unbalanced a commander in charge of their forces in Korea. Ironically, on hearing of his dismissal, MacArthur told his successor, General Ridgway, that he had heard from an eminent doctor that President Truman 'was suffering from malignant hypertension; that his affliction was characterized by bewilderment and confusion of thought'.

The Prima Donna

Inappropriate as it may seem to refer to a military commander as a prima donna there are occasions when no other expression will do. One such man was the distinguished French general Charles Denis Sauter Bourbaki, commander of Napoleon III's Guard Corps during the Franco-Prussian War of 1870-1. In the early part of the campaign the Guards had been kept in reserve and did not fire a shot in anger. Bourbaki clearly believed his men were being saved for something more important than the mere skirmishes which satisfied lesser commanders. His moment came at the great battle of Gravelotte-St. Privat, almost certainly the decisive battle of the war (see also p. 117).

Marshal Canrobert, commanding the right of the French line at St. Privat, saw his position crumbling under pressure from the Prussian Guard and the Saxons. He warned the French commander on his left, General Ladmirault, to send a messenger to call up Bourbaki and the Imperial Guard, which was in reserve. Bourbaki had been in position behind the French centre for most of the battle and had already refused to send a brigade to help out the hard-pressed General Frossard. When the French commander Bazaine questioned the decision, Bourbaki replied that he did not intend to commit the Guard too soon or in 'penny packets'. Incredibly Bazaine accepted this, saying that Bourbaki could suit himself how he used the reserve, and did not raise the matter again during the day. From early morning until 6.15 pm Bourbaki's proud guardsmen stayed well out of the fighting. It was at this time that the messengers from the French right arrived to ask him for help. Bourbaki was suspicious. Had the French line broken? He had no intention of wasting his reserve on a hopeless cause. The messengers convinced him that though the situation was dangerous it was not yet fatal and so he set off with them, leading a division of the Guards.

As he approached St. Privat, Bourbaki's worst fears were confirmed. The roads were filled with French soldiers fleeing from the front. He could not allow his men to be contaminated by these signs of defeat. He turned on the messengers and screamed at them, 'You promised me victory. Now you've got me involved in a rout. You had no right to do that! There was no need to make me leave my magnificent positions for this!' Furious, he turned his division about and led them off the battlefield. The sight of the Imperial Guard in retreat confirmed the views of men of both the 4th and 6th Corps that the day was lost. Panic spread down the line and a stampede set in. Bourbaki was now overtaken by events. His own guardsmen, seeing the panic behind them, now increased their pace and began to gallop away, deaf to his entreaties. Instead of receiving help General Ladmirault saw his whole front collapse thanks to Bourbaki's childish display. All he could do was withdraw as best he could under cover of darkness.

The exclusive club

The Russian attack on Kovel in July 1916 was a part of the otherwise successful 'Brusilov Offensive'. However, it was staged for the benefit of the Tsar and his group of military friends in the Guards Army, of whom the most incompetent was one General Alexander Bezobrazov. The Guards Army was an élite force which had been undergoing extensive training and was thought to contain the best troops in the whole Russian army. However, General Brusilov had no confidence in the Guards Army, in its plan to take Kovel, or in its commanders. Like Bezobrazov, the latter were all prima donnas – nominees of the Tsar chosen not for military ability but for amiability, camaraderie and good conversation at royal dinners.

The place chosen for the Guards to attack seemed to have been selected by the Germans rather than the Russians, for it was basically a swamp. The three causeways down which the guardsmen would have to advance were overlooked by German machine guns on both sides. The first assault was launched without any preparation whatsoever and was not supported by artillery fire. The Guards even had to cut through barbed wire before they could begin to advance. Through the raw courage of the guardsmen some gains were made, but at a cost of 30,000 of these élite troops.

Bezobrazov's colleague and the Tsar's bosom friend, the Grand Duke Paul, commanding the First Corps, was ordered to make a flank attack, but refused, claiming it was beneath his dignity and that of the Guards army to go skulking around the back. Instead he made a full frontal attack. Russia's two finest regiments, the Imperial Rifle Regiment and the Preobrazhensky Guards, marched in full splendour towards the German guns. So heavy were casualties on the causeway that many guardsmen chose instead to drop into the swamp alongside and wade through the weeds and slime. But their slow progress through the bog made them easy targets for the German guns. To add to their problems, German planes bombed and machine gunned them from above, while their own artillery, expecting them to be on the flanks and not in the centre, shelled them as well. Many drowned in the fetid water. Those who were rescued or taken back to the field hospital were then blown up when a fuse dump located next to the hospital exploded. Incredibly, the guards eventually won through and took the German trenches at a cost of 70 per cent casualties. The cavalry, who were supposed to consolidate these gains, refused to attack in such conditions and the attack was called off. The guards had to abandon the German trenches and march back down the causeway, still under German fire.

In all, the Guards Army – 'physically the finest animals in Europe' – had lost 55,000 men in this ridiculous operation. Anger was widespread. Professional Russian officers pointed out that if there had been cooperation between the prima donnas and the 'real' troops something might have been gained. Bezobrazov and the Grand-Duke Paul were both dismissed from their command, but soon restored to the Tsar's favour and dining-table.

The legend of George Armstrong Custer

Of all the commanders in this book few will be as well known to the general reader as George Armstrong Custer. And in playing the prima donna, Custer took lessons from no man. He was a master at getting out of tight scrapes and getting his own way by behaviour that would be considered outrageous in the nursery.

Everyone knows the Custer of legend, the star of several motion picture epics. He stands with golden hair blowing in the breeze and handsome face steadfastly set against a cobalt blue sky, while hordes of Indians ride round an ever-decreasing ring of blue, tipped with carbines defiantly spitting death. Under the flag of the immortal 7th Cavalry, Custer makes his last stand. And in the movies, of course, as the sun sets for the last time they die with their boots on, leaving behind at the fort their sobbing loved ones bedecked in yellow ribbons.

The truth is rather different. Lieutenant-Colonel Custer (he borrowed the rank of 'General' for appearance's sake) was one of the worst commanders ever to lead men in battle. His last stand was no epic, but a bungled operation involving a part of the 7th Cavalry in which Custer lost 211 men. The Indians did not scalp Custer at the battle of the Little Big Horn, as his golden curls were on a barber's floor somewhere. Perhaps expecting disaster Custer had a crewcut!

Custer would perhaps have been happier among the cavaliers of Prince Rupert's horse, rather than the hardbitten professionals of the US cavalry. He was an instinctive fighter, not a thinking officer. Obeying orders was for lesser men, as was studying his profession – Custer graduated from West Point 34th out of a class of 34. He had shown verve and daring in the Union cavalry during the Civil War, but little intelligence. Nor was modesty one of his virtues. After leading one successful charge he claimed, 'I challenge the annals of war to produce a more brilliant charge of cavalry.' He was a showman, who gained the local rank of Major-General of volunteers, and made the most of it by devising his own uniform, all velvet, glitter and gold. He loved publicity and won more renown than his achievements ever warranted. His egomania made him many enemies and few friends.

In 1867 he was court-martialled for an abysmal display during General Hancock's Indian reconnaissance, in which he disobeyed orders, forsook his mission to pursue his own ends, and treated his men cruelly, abandoning two of them to the tender mercy of the Indians. Reinstated the following year, he won notoriety for massacring 103

Custer's last stand at the Battle of Little Big Horn has become an American legend. The reality is less
heroic. Custer was a vainglorious egomaniac who brought defeat upon himself through disobedience
and ineptitude.

Cheyenne, including women and children, and earned from the Indians the epithet 'Squaw-Killer'.

When General Terry was sent to the Black Hills to subdue the Sioux in 1876, Custer's 7th Cavalry was included in his force. In fact, it was only Terry's influence in Washington that won Custer the chance, for he was then under a cloud, having quarrelled with President Grant, as well as Generals Sherman and Sheridan – strong opponents indeed. As usual Custer had been talking to the press and had spread allegations of corruption in high places that he could not justify.

In June 1876 General Terry's forces approached the main Indian encampment in the valley of the Little Big Horn River. Custer led one of three columns that were to converge on the Indian force. It was made quite clear to Custer that he was not to act alone, but he took no notice. He had been boasting that the 'Seventh' could whip the Indians alone and when it was suggested that he should take some Gatling guns he scornfully refused. Before he left camp Colonel Gibbons told him, 'Now, Custer, don't be greedy. Wait for us.' 'No, I won't', said Custer, ambiguously, as he galloped away.

Custer led his troops towards the Indian encampment, which contained perhaps 15,000 people, including at least 5,000–6,000 Sioux and Cheyenne warriors. Told by his own Crow scouts not to give away his position by making campfires, Custer ignored them. Naturally the smoke was soon spotted by the Indians. Now he had lost any chance of surprise. Seeing a group of Sioux riding away at speed Custer assumed the whole of the Indian force would panic when faced by a large body of US cavalry. His Indian scouts tried to tell him he was wrong. Again he would not listen. It is hardly surprising he was bottom of his class at West Point.

Custer divided his command into three parts, and decided to attack the camp from south, west and north. Major Reno launched a charge from the south with just 112 men, but was swept back by a massive Indian attack and had to take cover in the high bluffs. Here he was joined by Captain Benteen, whose attack, with 125 men, had been just as easily swept aside by the Sioux. But where was Custer?

Custer was preparing to attack across a stream at the northern end of the camp, three miles from the rest of the regiment and quite beyond help. But before he could do so the main body of the Sioux led by Chiefs Crazy Horse and Gall swept into sight, driving him back onto a slope. Surrounded, Custer and his men dismounted and enacted the now famous scene. His entire force, except for a single horse, was wiped out, and the bodies stripped and scalped. Custer himself, even without the golden hair, was recognized by the Indians and left unscalped.

Custer's generalship was so bad as to defy analysis. Few examples come to mind of so small a force attacking against such odds, though Gérard of Ridefort's performance at the Springs of Cresson in 1187 is perhaps comparable (see p. 63). Outnumbered ten to one, Custer was facing certain annihilation by disobeying orders and trying to storm the Indian camp. Then to divide his already tiny force after the Indians had spotted them was lunacy. Perhaps his drive for personal glory blinded him, as so often in the past, to his responsibilities. If anything, the career of George Custer should stand as a warning of the need for careful selection procedures in the military profession.

The Stupid

Much as one might like to find kinder ways of describing some military commanders, one is sometimes left struggling for any better word than 'stupid'. Examples of actions apparently prompted by stupidity are unfortunately very common in military history. They rarely cost the officer responsible anything more than his job, but often cost the lives of the troops under his command.

Forts . . .

In 1758, during the colonial wars in America between Britain and France, a force of British troops were advancing towards the French settlement at Fort Duquesne on the Ohio River. A certain Major Grant asked his commander, General Forbes, for permission to reconnoitre near the fort and perhaps take a few prisoners. With 800 men, including some Highlanders and some Virginians, he reached a point about half a mile from the fort. Here, according to General Forbes, Grant appears to have 'lost his wits'. First splitting up his force into small sections, which he then dispersed over a wide area, he sent an officer with an escort to sit in front of the fort and make sketches of the French defences. He compounded this folly by beating reveille loudly enough to wake the dead. The gates of the fort then swung open and French troops and Indians poured out, quickly overrunning the sketching party, killing 300 of Grant's men and capturing the Major himself.

A similar small expedition in India towards the end of the Mutiny in 1858 also gives us reasons to doubt the sense of its commander. The expedition in question has gone down in history as 'Walpole's Folly'. Sir Colin Campbell's choice of Brigadier Walpole to lead a strong force, including three Highland regiments, into Rohilkhand to pursue Bahadur Khan met with incredulity from those who knew the brigadier. He was seen as a brave man but an 'obstinate blunderer', and one officer described him as 'a great dolt'. Undeterred, Walpole proceeded to prove them all right. Approaching Fort Ruiya the British column was met by a British soldier who had just escaped from imprisonment in the fort. He told Walpole that the commander of the fort, Nirpat Singh, was overawed by the British troops and would only put up token resistance 'to save his honour', before surrendering, apparently having a garrison of just 200 men. Walpole dismissed this information as nonsense: he estimated the enemy garrison at 1,500. Refusing to scout around the fort, which was protected on two sides by dense jungle, he decided to assault the front, where its walls were high and thickly protected. As it turned out, and as he could have discovered by the most cursory reconnaissance, the wall at the rear was so low 'that a child would have had no trouble in climbing over it'.

It was quite true that Nirpat Singh intended to surrender, but faced with the incredible sight of British troops advancing in dense masses towards the strongest part of his walls, he had second thoughts. As the defenders opened fire over a hundred British soldiers were shot down, including the popular Highlander Brigadier Adrian Hope. Walpole now panicked and ordered an immediate retreat, being roundly cursed by his own men as a fool and coward. Ironically, that night Nirpat Singh evacuated the fort under cover of darkness, leaving it for Walpole to occupy the next day – a day too late as it turned out.

An uneasy alliance

Keeping an alliance together has never been easy, but during the Napoleonic Wars it was made harder by the incompetence of the Austrian and Russian military commanders. The Austrian General Weyrother won few friends among his Russian allies for his planning during the Alpine campaign of 1799. The Russian commander Marshal Suvorov, fighting the French in the Swiss Alps, had Weyrother on his staff as a geographical specialist. Weyrother advised him on a route from Altdorf to Schwyz, taking the Russian troops across the St. Gotthard Pass in appalling conditions and against strong French harassing attacks. Trusting the Austrian, Suvorov battled on only to find the road disappear against the sheer face of a mountain. Only the courage and extraordinary fighting qualities of the Russian soldiers enabled Suvorov to survive this blunder and retreat successfully. But Weyrother was not finished yet. In 1805 and with the same allies, he was responsible for planning joint strategy for the Austerlitz campaign. Incredible to relate, the Austrian planner overlooked the ten-day difference between the Gregorian calendar, used in the West, and the ancient Julian calendar, still used by the Russians. As a result the Russians under Field Marshal Kutuzov arrived ten days late for their rendezvous at the River Inn, throwing out the whole Allied plan.

. . . and more forts

The fall of Fort Douaumont at Verdun to a single German soldier in 1916 was the result of a crass error on the part of one man. The absent mindedness of General Chrétien – it can be argued – was to cost France the lives of 100,000 men. Douaumont, reputed to be the strongest fort in the world, had been stripped of manpower in the months before the German assault on Verdun, but in February 1916 Chrétien was ordered to reoccupy the fort and defend it to the last man. As Chrétien was going on leave he should have passed the message to his successor, but he simply forgot. As a result, at the start of the great German assault on Verdun, Fort Douaumont was manned by just 56 gunners.

'There are objectives within reach . . . for the retention of which the French General Staff would be forced to throw in every man they have. If they do so the forces of France will be bled to death, since there can be no question of a voluntary withdrawal, whether or not we reach our goal . . . The objectives of which I speak are Belfort and Verdun . . . The preference must be given to Verdun.'

General Erich von Falkenhayn, December 1915

Falkenhayn's decision to introduce attrition into warfare, namely the deliberate sacrifice of large numbers of your own men on the assumption that the enemy will lose even more than you do, was a logical outgrowth of the idea of total war. In such a struggle every resource must be used, including manpower. However, it was also a tacit admission that current tactical and strategic skills were inadequate to deal with the military stalemate.

On 25 February the 24th Brandenburg Regiment took up positions half a mile from the fort. A section of Pioneers, led by a Sergeant Kunze, had been instructed to clear away barbed wire obstacles and positioned themselves in the moat surrounding the fort. Kunze formed his men into a human pyramid so that he could climb into one of the gun embrasures. Finding no one there, Kunze began exploring the long corridors, arresting four French gunners in the process. He lost these prisoners, but stumbled into a room where a lecture was taking place and arrested twenty more Frenchmen. Kunze locked them in the room, and made his way to the officers' mess, where he sat down and had breakfast. The capture of the fort was completed by three German officers, who had been apprised of the situation by Kunze's colleagues in the moat.

News of the capture of Fort Douaumont was greeted with horror in Britain and France, while the Germans were triumphant. To cover the shame of its fall the French claimed it had cost the Germans thousands of lives and no attempt was made to contradict this by the German authorities. It remained in German hands for six months until recaptured at terrible cost by French colonial troops in October 1916.

A generation later the French were still having trouble with forts. Due to a ridiculous mix-up in May 1940 the French found themselves locked out of their own Maginot Line. When the commander of IX Corps, General Martin, was ordered to advance he locked his concrete bunkers and left the key with a colleague, on the staff of the 53rd Division. But when this division was ordered south nobody remembered that they held Martin's keys. When Martin's forces were driven back by the Germans the General tried to reoccupy his defences only to find that he had not got the keys. With the Germans snapping at his heels, Martin had to get his engineers to break into the bunkers. In fact, he might have saved himself the trouble as his corps was soon overrun by Rommel's tanks.

'It appears certain that, after one serious defeat, they [the Boers] would be too deficient in discipline and organization to make any further real stand.'

Military Notes on the Dutch Republics by the War Office Intelligence Department, 1899

One might have expected that after the experience of the First Boer War the Intelligence Department would give more appropriate advice than this to its commanders going out to South Africa.

Sir Reverse Buller

In any discussion of military stupidity the career of General Sir Redvers Buller assumes a prominent place. Of Buller it has been said that he was an excellent major, a good colonel and an atrocious general. Unfortunately, it is as a general that we must judge him. In the 1899 military manoeuvres at Aldershot, Buller, who had not commanded troops for twelve years, ordered his men to make a full frontal assault against an equal number of defenders. His men had just finished a fourteen-mile march and were in no condition for attacking anybody. Not surprisingly the manoeuvres had to be abandoned. Buller's conduct of the manoeuvres was comically inept: no trenches were dug for fear of damaging the countryside, no man was allowed to dive for cover lest he damage his uniform, and the soldiers eventually had to resort to volley firing at each other in the open and at ranges of less than a hundred metres. To cap it all, the manoeuvres could only take place between 9 am and 5 pm so that they did not interfere with the officers' social diaries. This was how Buller and his comrades prepared Britain for the rigours of twentieth-century warfare. In fact, they did not have long to wait before putting their training to the test. Within months, war had broken out against the Boers in South Africa and Sir Redvers Buller was to command the British expeditionary force.

Buller's arrival in South Africa in 1899 brought a new level of ineptitude to the British war effort. Some might have thought Sir George White was doing well – or badly – enough on his own, but with Buller came generals who seemed to lack not only military competence but common sense as well.

'The Boers are not like the Sudanese, who stood up to a fair fight. They are always running away on their little ponies.'

General Herbert Kitchener, South Africa, 1900

The British idea of 'fair play' consisted of the enemy standing in the open and being shot down by British rifles and machine guns. Hiding behind rocks and trees was not playing the game.

Within days Buller's master-plan to relieve Ladysmith was in tatters. Commanders such as Gatacre, Methuen, Hart, Long and Warren had seen to that. General William Gatacre – 'Backacher' to his men – set out to capture the strategic railway junction at Stormberg by a surprise dawn attack. This entailed his force of 2,700 troops undertaking a difficult night march, during which they completely lost their way. It appears that the one man who knew the terrain and could have led them safely had been left behind. As dawn broke the British troops were totally lost, stranded at the base of a steep cliff on top of which the Boers were quietly drinking their morning coffee. When they saw the British below them the Boers opened fire, while Gatacre's men made brave but unavailing attempts to climb the almost sheer rock face. Seeing his men falling back in confusion, Gatacre ordered a retreat which degenerated into a mad scramble as a Boer commando tried to cut them off. On reaching safety, Gatacre was congratulating himself on having suffered only 90 casualties, when the realization dawned on him that he had forgotten over 600 men who had received no orders to retreat and were still on the slopes. Surrounded by the Boers hours after the rest of the army had abandoned them, they had no option but to surrender. 'Better luck next time,' Buller telegraphed to Gatacre.

Not to be outdone by Gatacre, General Lord Methuen bombarded a hill near Magersfontein on the assumption that the Boers were entrenched at the top. In fact, they were entrenched at the bottom and watched the pyrotechnic display like amused youngsters at a fireworks party. Assuming that the Boers were thoroughly rattled by the big guns, Methuen now prepared a night march (shades of Stormberg). Brigadier Wauchope was to lead 3,500 men from the Highland Brigade in the absolute darkness and torrential rain of a moonless night. More through luck than judgement, the brigade arrived at the exact point Methuen had chosen, from which they could dimly see the outline of Magersfontein Hill. What Methuen did not know was the exact location of the Boers, and so unwittingly he was sending Wauchope into a trap.

As dawn began to break the Boers caught sight of Wauchope's Highlanders, still marching shoulder-to-shoulder, about 400 yards away. They immediately poured heavy fire into the quivering ranks of the Black Watch. One man described it as if 'someone had pressed a button and turned on a million electric lights'. Some men took to their heels, but most fell to the ground, taking cover behind anthills or bushes. As the heat of the day increased the Highlanders found that they were trapped, with the sun blistering the backs of their knees and the ants driving them mad with their bites. Yet to stand up was to die in an instant from the bullet of a Boer sharpshooter. Methuen was little more than a spectator at the scene. With his men trapped out in the open and under heavy fire he seemed paralysed by indecision. At last the men of the Highland Light Infantry gave way to panic and stood up to flee, many of them being shot in the back as they ran. The body of their commander, the popular Andy Wauchope, was found near the Boer trenches. Altogether British losses in killed and wounded amounted to 902 men.

After his personal defeat at Colenso (see p. 37), Sir Redvers Buller was reinforced by a new division under Sir Charles Warren, probably the most preposterously incompetent general to serve in the Boer War. Warren, curiously, had earlier been the police commissioner who had failed to trap Jack the Ripper. In his latest incarnation, as a 59-year-old general who had been on the retired list for a year, Warren brought to South Africa some military theories that were quite ridiculous. He had told Lord Wolesley, the Commander-in-Chief, that the way to beat the Boers was 'either by sweeping over them with very long lines of infantry attacking simultaneously', or 'by pounding away at them with artillery till they quailed'. Nevertheless, Warren was promoted to Lieutenant-General and packed off to add to the confusion on the Cape. In February 1900, while the fighting raged around Hussar Hill, Warren kept his men entertained and his officers appalled by bathing in public. What Buller thought when he rode up and found his colleague splashing about in the water instead of directing his troops is not recorded. Warren also believed in joining the men in spanning the oxen teams and hauling on ropes like a common soldier. Such eccentricities might not have mattered had he shown good sense when it really counted. During the crossing of the Tugela at Trickhardt's Drift, Warren spent 26 hours supervising the transfer of his personal baggage across the river. When he first reached the river a mere 600 Boer defenders were in position, but during the time it took Warren's baggage to cross a further 6,000 had moved into position west of Spion Kop to block his advance.

Buller was certainly worried by Warren's 'fads and fancies' and considered him 'aimless and irresolute'. Yet he took no action against Warren and appointed him commander of the British forces at that most infamous of Boer War battles, Spion Kop. Buller and Warren agreed that the isolated hill of Spion Kop was the key to the situation since it would command the Boer positions. However, neither ordered a proper reconnaissance of the area, and simply ordered troops to occupy the top, though what they were to do after that was never decided. Warren immediately selected General Talbot-Coke – who had only just arrived in South Africa and was suffering from a leg injury – to lead the attack. When Buller questioned this, Warren compromised and replaced Talbot-Coke with a 55-year-old, General Woodgate, of whom Buller said that he had two good legs but no head.

In planning the assault it never occurred to Warren to send up machine guns to increase the British firepower, or a field telegraph unit to keep him in touch with the men on the hill. Sandbags had been prepared to fortify the top, but they were forgotten as well, and just 20 picks and shovels were taken to entrench 2,000 men.

In thick fog the assault troops, mainly from the Lancashire Brigade, set off up the steep-sided hill, while the rest of the army – nearly 20,000 men – simply stood and watched. As the younger and fitter men pulled ahead, Woodgate fell further behind until he was being almost carried. His men, told to empty their rifles and use the bayonet alone – usually a signal for disaster – reached what they took to be the summit and made attempts to entrench themselves in solid rock. As visibility began to improve they discovered they were not at the top at all, but on a sort of plateau some way from the actual peak. Higher up there were three better positions which the British troops could have occupied, but the Boers raced to occupy them instead and opened fire on Woodgate's men from three sides. Trapped on a plateau no more than a quarter of a mile square and without proper cover – for their attempts to dig trenches were pitifully inadequate – the British troops were massacred by Boer sharpshooters. Some troops fought with great heroism while others simply surrendered. For all that Warren knew of what was going on his men might as well have been on the dark side of the moon.

When Winston Churchill, who was present as a war correspondent, tried to convince Warren that his men were trapped on Spion Kop, the general flew into a rage and ordered his arrest. Reinforcements were desperately needed, as well as artillery support. Meanwhile on the kop Woodgate had been killed and Colonel Thorneycroft had taken over command. But for nine hours no message reached the new commander from Warren, and eventually he ordered a withdrawal from the hill. Ironically, it was at this moment that Warren decided to occupy the hill in force. As the survivors made their way down the rocky slopes they met fresh troops climbing up to meet them. But it was too late. The Boers were now masters of Spion Kop and Warren was forced to call off the entire operation.

As one of the participants at Spion Kop wrote, 'There was no plan except that we were to take the hill and stay there. Some 1,700 men were to assault a hill 1,740 feet high in the centre of the Boer position, and the rest of Buller's 20,000 men were to look on and do nothing.'

The battle was a saga of blunders and missed opportunities from which none of the commanders emerged with any credit. Warren's actions – or occasionally inactions – displayed an extraordinary blend of naïvety and incompetence. Such faith was placed in the bayonet as a weapon that the attackers did not take with them the necessary artillery, machine guns, entrenching tools, food or water to enable them to hold the hill for any length of time. Buller wrote to his wife, 'We were fighting all last week, but old Warren is a duffer and lost me a good chance.' Yet Buller was equally to blame for leaving Warren in charge of the battle. From now on even Buller's most faithful supporters, the common soldiers, began to speak of him as 'Sir Reverse Buller'.

Uncle Jo's Buddy

Marshal Semyon Mikhailovich Budenny's claim to being the most incompetent commander in this whole book is a strong one. So elevated did he become in the military profession in relation to his ability that oxygen starvation must be suspected. Against the

Germans in 1941 he suffered probably the biggest defeat ever suffered by any general in military history. Even to his own men he was the man 'with the very large moustache and the very small brain'. His meteoric rise to a position where he could do such damage stemmed from his friendship with Joseph Stalin and his obvious lack of ability. For Stalin wiped out all those Red Army officers who might have posed even the slightest threat to his dictatorship. The fact that Budenny survived and prospered speaks volumes.

A sergeant major in the Imperial Russian cavalry in 1917, with the Revolution Budenny overnight became commander of the First Soviet Cavalry Army. This was not, one might confidently suppose, through any military ability, but through being a close supporter of Stalin. Budenny's military theories might have been gleaned from previously unsuspected texts by Attila the Hun and Genghis Khan, consisting of charging at everything and everybody with the utmost gusto. Flamboyant, hard-drinking, but essentially simple, Budenny at least added colour to the drab years of early Soviet history. However, at a time when German generals were developing modern mechanized warfare, Budenny's ideas were rooted in the past. During the purge of the Red Army, in which Stalin eliminated 3 marshals, 13 army commanders and 400 senior generals, Budenny won promotion as Deputy Commissioner for Defence. How seriously he took this position can be illustrated by the story of a party held to honour his visit to inspect the military area of Bessarabia. After everyone had eaten and drunk their fill, a canvas screen was removed revealing a huge vat filled to a depth of five feet with red wine. Swimming and splashing inside the vat were naked girls. Budenny, swiftly stripping off his clothes, jumped into the vat, mustachios and all, followed by several of his aides. Frustrated at not being able to get in, one of his other guests fired a long burst at the vat with his automatic weapon, wounding several of the girls and flooding the floor with the wine. Bellowing with laughter, Budenny carried off one of the maidens, none the worse for this brush with danger.

Already senile at 58, Budenny commanded the Russian forces in the Ukraine and Bessarabia against the Germans in 1941. Even though he outnumbered the Germans three or four to one in men and tanks, Budenny could do little as Rundstedt and Kleist literally ran rings round him. Between July and September 1941 Budenny's incompetence cost the Russians 1,500,000 men. Even when he was removed from his command he was awarded the title 'Hero of the Soviet Union', a remarkable example of Stalin remaining loyal to his friends.

> *'War is a contest between two human intelligences rather than between two bodies of armed men. Ligny and Waterloo were won by sheer power of brain.'*
>
> **Introductory Lecture at Staff College, 1896**
>
> Ironically, the products of the Staff College were noted for anything but their 'power of brain', as the Second Boer War was soon to show.

THE BATTLE OF KARANSEBES
(1788)

The Austrian Emperor Joseph II had lived for too long in the shadow of the Prussian king, Frederick the Great, and by 1788 he was determined to achieve a great military success for himself. Although in a pitiful state of health, with a weak heart, varicose veins, and a dry, hacking cough, he decided to lead an army against the Turks in Transylvania. Driven by a desperate urge to succeed and calling himself the 'avenger of mankind', he declared he was going to 'cleanse the world of a race of barbarians'.

The campaign began badly when Joseph decided, against the best local advice, to camp in a malarial area near Belgrade. In the space of six months 172,000 of his troops fell ill with the disease and 33,000 died. Joseph worked frenziedly to prepare his troops for the campaign ahead. Rarely able to keep down any food himself and drinking just a little water, he resembled a living corpse rather than a man. Then news arrived that the Turkish army, under the Grand Vizier, was moving towards him. Joseph immediately decided to take half his force and seek battle. Near the town of Karansebes he got his wish – but not quite in the way he expected.

With the Turkish army still some distance away, the Austrian columns were marching in good order, flanked by regiments of hussars. As night fell the army crossed a bridge near the town of Karansebes, watched by a number of travelling Wallachian peasants. Apparently some of the hussars stopped to buy liquor from the pedlars. An incident occurred when some of the infantry, tired from a day of footslogging, also left their ranks to buy some, only to be driven away by the hussars. The footsoldiers, furious at the arrogant cavalrymen, fired some shots in the air and tried to frighten them by yelling 'Turci! Turci!', pretending they were about to be attacked. Joining in what seemed to be a bit of fun, the now drunken hussars also shouted 'Turci!' and let off some shots as well.

The result was startling. The rear columns of the army were still approaching the bridge and, hearing firing and shouts, they began to panic, firing at each other in the darkness. Officers rushed up and down the columns shouting 'Halt!', which to panic-stricken ears sounded like 'Allah!' To the Austrian troops it seemed that the whole Turkish army had ambushed them. The baggage handlers and transport workers at the rear of the army were seized with terror and tried to drive their wagons through the massed troops ahead, knocking soldiers in all directions and spilling many into the waters of the river. Then, with a great roar, thousands of men began to stampede in the darkness.

The first the Emperor, sick and prostrate in an open carriage, knew of what was happening was when a flood of men, horses and wagons swept his carriage off the road and threw him into the river. Bravely mounting a horse he tried to rally his troops, but by now there was no hope. Heavy fighting had broken out on both sides of the bridge, and at the rear wagons were overturned, baggage lost and dozens of cannon abandoned. Everywhere the cry went up, 'The Turks are here; all is lost; save yourselves.' By first light it was apparent that the Austrians had suffered a major disaster, having lost over ten thousand men killed or wounded by their own colleagues.

RETREAT FROM KABUL (1842)

The retreat from Kabul during the First Afghan War is one of the epic stories of British military history. Of 16,000 men, women and children who set off through the Khyber Pass in winter, only one European and a few Indian soldiers survived to tell the tale. It was a catastrophe, but it was an avoidable one. Mistakes were made throughout, but none greater than the choice of commander, Major-General William Elphinstone, who was by common consent physically incapable of taking on such a difficult assignment. Never was a more unsuitable appointment made, and for this the Governor-General of India, Lord Auckland, was totally to blame. Elphinstone's physical condition was widely known before he was selected for a command that would have tested the health of a much younger and fitter man. To this extent Elphinstone was as much a victim of military incompetence as he was guilty of it himself.

The British retreat from Kabul in 1842. The 16,000 men, women and children who set out to march to India met a savage fate at the hands of Afghan tribesmen in the icy wastes of the Khyber Pass.

To attempt to describe the state of Elphinstone's health in 1842 one needs to ransack a shelf of medical dictionaries. He provided enough ailments for a whole convention of doctors. Nearly sixty, flatulent and incontinent, and with developing senility, the unfortunate general had one arm in a sling and had such bad gout and rheumatism that his legs were crippled. He could neither walk nor ride and had to be carried everywhere in a litter. To expose such a cripple to the extremes of heat and cold to be found in Afghanistan was cruel in the extreme. Elphinstone had tried to refuse the appointment, but Auckland had insisted, and the gentle old invalid had given way, as he was to do throughout the days ahead, with no more than a feeble, self-effacing smile.

The decision of Lord Auckland to occupy Afghanistan by military force and supplant the Amir of Kabul, Dost Mohammed, with the feeble Shah Suja was a military blunder in its own right. In 1839 British troops under General Sir John Keane drove Dost Mohammed into hiding and installed Shah Suja in Kabul. General Sir Willoughby Cotton was left with 5,000 men in the capital and General William Nott with a similar force at Kandahar. Other garrisons were established along the route back to India. But the

Afghans never willingly accepted the British occupation. With the army settled in Afghanistan, numerous wives and families of officers and even the sepoys were brought to live in a new cantonment in Kabul. The Afghans took this as a sign that the British had a long occupation in mind and plans were laid for a general rising.

Blissfully unaware of this, the soldiers in Kabul built their cantonment without giving a thought to how it could best be defended if the need arose. They chose an area of low, swampy land surrounded by hills and forts, all of which were occupied by the Afghans. In the event of a rising the whole cantonment would be open to bombardment from the heights. The cantonment perimeter, nearly two miles in circumference, was far too extensive to be properly defended, while the commissariat stores were placed four hundred yards away outside the walls. This was too much for the Commissariat Officer, who demanded they be placed inside, but was told by General Cotton that he was too busy building barracks to worry about the stores. When the rising came, it did not take Napoleonic insights to persuade the Afghans that once they had captured the British stores they would soon be able to starve the foolish infidels into surrender.

Elphinstone's appointment as Cotton's replacement set the seal on a woeful catalogue of tactical blunders. A colleague described Elphinstone unkindly but accurately as the most incompetent general in the British army. He had last seen action at Waterloo, where he had done well, but had for long years been retired on half-pay, before travelling to India to command a division of the Bengal army. Here his health, always poor, had collapsed completely, rendering him little more than an invalid. That this did not disqualify him for his appointment speaks volumes for the complacency of the Indian authorities towards the occupation of Afghanistan. To make matters worse, they saddled the gentle 'Elphy Bey' with the efficient but highly abrasive Brigadier John Shelton as his second-in-command. Shelton was a brute. His animal courage was combined with a morose personality, making him a terrible man to cross. He took an instant dislike to his commander, having nothing but contempt for his weakness. Elphinstone, for his part, disliked and feared the brigadier. On their arrival in Kabul they were assured by the retiring commander, General Cotton, 'You will have nothing to do here, all is peace.'

Cotton was never more wrong. Once Elphinstone took over, the underlying unease with the British occupation became more apparent. In October 1841, General Sale's brigade marching from Kabul to India was attacked by Afghans and had to take shelter in the fortress of Jalalabad. Meanwhile, Elphinstone's health deteriorated further; gout and rheumatism were now affecting most parts of his body. Lord Auckland decided at last to send a replacement, but by this time it was too late. His choice, General Nott, was besieged in Kandahar and could not reach Kabul. It was up to Elphinstone now.

Offences against British personnel were becoming a daily event, with soldiers being shot even inside the cantonment. Going out alone was often fatal. When the Residency was stormed and the British Resident, Sir Alexander Burnes, murdered with his staff there was no turning back for the rebels, who had taken Akbar Khan, son of Dost Mohammed, as their leader. The British were soon besieged in their cantonment and, with the loss of the commissariat stores, reduced to three days' supply of food. Inside the cantonment the British troops were furious at the sight of the Afghans pillaging the stores and making off with their food and liquor. The men pressed Elphinstone to let them take the offensive in Kabul, but Elphinstone could not make up his mind what to do. He even asked the advice of the British Envoy, Sir William Macnaghten, who duly told him to think again. All Elphinstone could think of was his health. Misfortune befell him when, feeling fit enough to mount a horse for a while, he promptly fell off and hurt his leg when the horse inconsiderately trod on it. By now Elphinstone's mind was seriously disturbed. As a way of accounting for the desperate situation they were in, he now decided that his men were running out of ammunition. There was, in fact, enough ammunition to withstand a twelve-month siege. Poor Elphinstone next wrote to the Envoy, standing a few paces away from him but barely on speaking terms, to warn him that in the absence of adequate ammunition some arrangement might need to be made with the enemy.

Macnaghten was no soldier, but he could recognize military incompetence when he saw it. He therefore sent a message to General Sale at Jalalabad, asking him to return to Kabul post haste to take over the command from Elphinstone.

However, even though the formidable Lady Sale and her daughter were in Kabul, the good general was unable to return without risking a disaster. Kabul was cut off and the garrison would have to fend for itself.

Meanwhile Shelton was treating Elphinstone abominably. The old general wrote in his report, 'I regret to be obliged to disclose that I did not receive from Brigadier Shelton that cordial co-operation and advice I had a right to expect; on the contrary . . . he invariably found fault with all that was done and canvassed and condemned all orders . . .' In fact, Shelton refused to tell Elphinstone anything at all about the military situation. All the old man knew was what he could eke out of conversations with the civilians. Elphinstone's 'Councils of War' were mere charades, at which anyone who wanted to turned up and gave what advice they liked. Junior officers argued with senior ones and civilians lectured soldiers. The boorish Shelton, meanwhile, brought his bedding to each meeting, unrolled it on the floor and lay upon it pretending to be asleep and snoring loudly. Elphinstone tolerated more contempt than any officer on record, before or since. When anyone took his side and tried to upbraid Shelton, the brigadier replied, 'I will sneer at him! I like to sneer at him!'

Although Shelton was a boor he was no coward. He twice led out men to clear the rebels from the hills overlooking the cantonment. On the first occasion, the 44th Regiment were charged by Afghan cavalry. They held their fire until the horsemen were no more than twenty paces away and then fired an explosive volley. When the smoke cleared it could be seen that not one single Afghan, horse or man, had been hit. The 44th promptly turned and ran. On the second occasion, Shelton's skill in deploying his men left much to be desired. He arrayed them in squares, one behind the other, with his cavalry behind both squares in the rear. The Afghans, expert marksmen that they were, now found themselves offered a concentrated target unknown to such tribesmen. British casualties were heavy, several men falling from each bullet that hit. Shelton, for some reason, had taken just one cannon, even though Indian Army regulations forbade the use of less than two in any action. At first this gun kept the Afghans at long distance, but when it became too hot to operate the Afghans closed in to short range. They laughed as the British muskets, misfired or badly aimed, allowed them to walk up to almost point-blank range. So feeble was the British musketry in comparison with the effectiveness of the Afghan that it comes as no surprise to read that British officers eventually resorted to picking up stones and pelting the rebels to make them take cover. Refusing to retreat, Shelton stubbornly watched his men being picked off one by one. He was struck on five separate occasions by spent bullets, but it is doubtful if he would have returned alive to the cantonment had his men not decided to run away. He was forced to follow them as they fled, arriving in time to meet a piqued Elphinstone, who told him that he had tried to rally the men but that when he had given them an order they had all looked the other way. Lady Sale, watching the whole fiasco, noted that from what she could see the British infantry had been driven back not by Afghan tribesmen, but by the tradesmen and artisans of Kabul itself.

The situation continued to degenerate. Elphinstone, to add to his range of ailments, now managed to get shot through the buttock. In an attempt to avert a catastrophe, Macnaghten agreed to meet Akbar Khan to discuss the situation. But Akbar had him murdered and his head and torso paraded throughout Kabul. Even now Elphinstone did not react. Still he could not make up his mind what to do. On a promise of friendship and protection, he agreed to evacuate the cantonment in mid-winter and march by way of the Khyber Pass back to India. With snow a foot thick, some 16,000 men, women and children marched to their death on the instructions of an invalid whose mind had collapsed under the strain of command. No sooner had they left the cantonment than they were attacked, and a constant attrition was kept up as they marched, with tribesmen from every area they passed through joining in the killing. It was said that even children hacked down the helpless soldiers and butchered them in the snow. Elphinstone died early on, and the rest over a period of a week. Only one European, Surgeon-Major William Brydon, reached safety at Jalalabad. Akbar Khan followed up his success by besieging Jalalabad, but General Sale marched from the city and routed him. Britain enacted due revenge for the massacre of the Kabul garrison when General Pollock became the first commander to force the Khyber Pass. With Nott he marched into Kabul and burned the great bazaar.

Elphinstone did eventually reach Jalalabad. When Akbar Khan heard of the general's death he had his body, wrapped in felt blankets and surrounded by scented leaves, crated up in a wooden frame and sent off under guard to Jalalabad. Even now he was not allowed to rest in peace. His escort was ambushed by tribesmen, the coffin cracked open, the body stripped and pelted with stones. Only the threat of what Akbar might do to them stopped them from burning it. A new expedition was sent to recover the body and it was eventually buried with full military honours.

THE KILLING FIELDS OF PARAGUAY
(1862–70)

It must be unique in military history for a commander to prolong a struggle so that his army, and virtually the entire population of his country, are wiped out altogether. But this was the case with the president of Paraguay between 1862 and 1870, Francisco Solano López – once described as 'a tidal wave of human flesh . . . a veritable mastodon, with a pear-shaped face . . . and heavy pendulous jowls'. López was a physically repulsive megalomaniac who hoped to be a great commander, the Napoleon of South America, and even kept one arm inside his jacket to ape the French Emperor. He had neither friends nor rivals in Paraguay, having shot them all, but he had a formidable accomplice in his mistress, an Irishwoman named Eliza Lynch, possessed of a personality almost as dreadful as López himself. Indeed it was Madame Lynch, who claimed to be the niece of one of Nelson's officers at Trafalgar, who fanned the flames of López's vanity, filling his mind with ideas of military greatness. Throughout his wars she was always at his side urging him onwards.

In Paraguay's war against the Triple Alliance of Argentina, Brazil and Uruguay, Lopez's motto was 'Victory or Death', though 'Victory and Death' would have been a more appropriate rendering. López trained his men so hard that many died, and no Paraguayan was ever allowed to surrender. The war took a terrible toll in human life, Paraguay's population being reduced from 1,337,000 to just 221,000, of whom only 28,000 were adult males.

To celebrate the expected victory of the Paraguayan general Estigarriba, Eliza organized a grand ball in the capital, Asunción. Everyone was commanded to appear in their richest clothes and the ladies to wear all their jewels. In the ballroom the orchestra was striking the first chords to welcome the presidential party when a rider brought López news of his army's surrender. The President lost all self-control, calling down curses on Estigarriba and demanding that he be executed on the spot – the general wisely chose not to return to Asunción and his family was massacred in his stead. Eliza went to the ball alone, placed an oil painting of López on his empty throne and everyone obediently bowed to his likeness throughout the evening.

López decided there was nothing for it but to take command himself and appoint Eliza Lynch as his Regent in Asunción. With his mind teetering on the brink of insanity, the 'South American Napoleon' began the campaign which would make his name famous throughout the world – though not quite for the reasons he hoped. Battles were fought without tactical skills and men were sent into battle to kill or be killed. López's abilities as a general were minimal. At the First battle of Tuyuty the Paraguayan forces launched an attack through woods into the face of massed Brazilian artillery firing grapeshot. They continued attacking until they were wiped out, losing 10,000-12,000 men, no quarter being asked or given. But López did not lose all his battles, and when he won the casualties were equally heavy on the other side. At the battle of Curupaity, the Paraguayans lost just 50 men in inflicting 9,000 casualties on their Argentinian and Brazilian attackers, who attempted repeated assaults on a heavily defended position. When ordered to retreat, the Argentinians displayed such contempt for their enemies that they refused to turn round and retired by walking backwards. In such a merciless war one is hardly surprised to learn that the Paraguayans tossed wounded and dead Argentinians into a nearby lagoon to be eaten by crocodiles.

This dreadful war of attrition was punctuated by moments of grim farce, as when López was nearly wounded by a shell from a Brazilian ironclad and decided that it had been aimed at him personally. In a rage he ordered his naval commander, a feeble

The Paraguayan dictator Francisco Solano López aspired to be the Napoleon of South America. But his brutal megalomania condemned his country to a war so devastating that it reduced its male population by nine-tenths.

old officer named Meza, to destroy the entire Brazilian fleet of modern ironclads with the few rafts that the Paraguayans had at their disposal. Fortunately Meza was so prostrate from age and ague that the command was taken by an English engineer named Watts. The Englishman, true to the spirit of Nelson, led boarding parties of barefoot native troops, who swung from the ropes with the agility of monkeys onto the decks of the Brazilian boats. The Brazilian sailors set up a howl and rushed below decks, pulling down the hatches. But the Paraguayans, drunk with their success, seemed not to know what to do next and marched up and down the ironclads laughing and banging drums. Suddenly, the hatches flew open and the Brazilian sailors poured back on deck and routed the Paraguayans, some of whom simply jumped overboard and swam ashore laughing. The Brazilians now took the opportunity of wiping out what was left of López's fleet. Brave Captain Meza was brought ashore, mortally wounded by a bullet in his lung. López told him to hurry up and die or he would have him executed for incompetence.

With the military situation growing ever more hopeless, López's moods swung from optimism to pessimism with alarming regularity. Fearing attempts on his life, he organized a spying system in the army whereby every third man was empowered to spy on his neighbours and shoot anyone, including officers, who showed any sign of cowardice. Thus Paraguayan armies marched into battle looking both ways and guarding their backs as well as their fronts. Distrust was the basis of López's system of dictatorship. During the siege of the Paraguayan fort of Humaita, 400 defenders under Colonel Martinez faced up to 30,000 Brazilian, Argentinian and Uruguayan troops. The plight of the defenders was so desperate they had been obliged to eat their horses, were surviving on roots and had little water left. They sent back to López asking for help, but he told them to hold on for five more days and then retreat. Martinez was so shocked at this that he tried to blow his brains out, but missed and merely shot one eye out. When the Allied forces launched an all-out attack on the fort the one-eyed commander and his men drove them off. But it was their last throw, and after a truly heroic defence they surrendered. Escaping from his captors, the commander, with a few men, somehow managed to fight his way through jungle and swamp to reach López with the news of their capitulation. It would have been better for him if he had stayed a prisoner, for an infuriated López had him and his men tortured and shot, along with their families.

López's mental state was hardly improved by his mother's revelation that he was a bastard and had no right to be president (his father had been president before him). López then announced that he would bring down the whole world around his ears. Ordering the entire population of Asunción to evacuate the city, he marched them off into the jungle, seeking a nineteenth-century 'Killing Fields'. Behind him he left as monuments to his insanity vast unfinished edifices: a palace worthy of an Emperor of South America, and an opera house to rival Paris or Milan. He considered calling on his people to commit mass suicide so as to rob his enemies of their victory. He even sentenced the head of his bodyguard to death, because

the unfortunate man had been unaware of a plot against López's life. In his fevered state he mixed religious fantasy with sexual indulgence, insisting that he should be made a saint and then ordering men and women to copulate and indulge in a variety of perversions in his presence. Rather than attempting to control him Eliza Lynch egged him on and even acted as his procuress.

Not to be outdone by her presidential lover, Eliza organized Paraguayan women into 'Amazon' regiments, equipping them with weapons and teaching them to fight with the lance. These women fought in a number of Lopez's battles. On one occasion, led by Eliza – riding bareback on a horse – they actually succeeded in driving the Allied troops back. The episode was not without its misfortunes, however, as Eliza's horse dropped dead and rolled on top of her, causing her extensive injuries.

Suffering defeat at the battle of Pirebebuy, López now ordered the national treasure to be thrown over a cliff into deep jungle, and then for good measure had all the witnesses thrown over the cliff as well in the interests of secrecy. Having failed to join the pantheon of the great commmanders, López had himself declared a Saint of the Christian Church. The matter was put to the bishops of Paraguay and the 23 who did not agree were shot. The ceremony then proceeded, Francisco's vast shape was anointed and blessed by the remnants of the church hierarchy, and that date entered into the Christian calendar. Duly canonized, López's first act was to have his 70-year-old mother publicly flogged. He then ordered the minting of a new medal for the whole population of Paraguay, what was left of it, except for his mother, who was to be executed instead.

Francisco Solano López, this latter-day Caligula, had just finished signing his mother's death warrant when his camp was attacked by Brazilian soldiers. Fighting to the last, he tried to run off but was hindered by his enormous bulk. A Brazilian soldier known as Chico Diablo threw a spear at López and wounded him in the stomach. Floundering about like a beached whale, López fired several shots from his revolver before expiring, while the serene Eliza was rescued from the fury of her own vengeful Amazons by an elegant Brazilian general, with whom she left the unsavoury scene.

THE BATTLE OF COLENSO (1899)

Sir Redvers Buller's first task as commander of British forces in South Africa in 1899 was to relieve the besieged garrisons in Ladysmith, Mafeking and Kimberley and then to beat the armies of the Transvaal and the Orange Free State. By the time he was replaced by Field Marshal Roberts he had made such a mess of it that he had ruined his own career and undermined Britain's military reputation in the eyes of the world.

Buller's army was divided into three parts: Lord Methuen with a division failed dismally to relieve Kimberley, General Gateacre completely mismanaged his brigade, while Buller himself with the largest force failed to relieve Ladysmith. Few people either in Britain or South Africa could have expected all three generals to fail so badly, or that all of them would make such elementary mistakes.

Between Buller and Ladysmith lay a Boer army of unknown strength. It was commanded by General Louis Botha and was entrenched among the hills north of the Tugela River. It was Buller's job to get past Botha, yet as he himself freely admitted,

the Boer position was a strong one. In a telegraph to the Secretary of State for War, Lord Lansdowne, on 12 December, he ruled out the idea of a frontal assault as likely to prove too costly. But having said this he immediately changed his mind.

Buller next signalled by heliograph to General White in Ladysmith that he was planning to attack the Boers on 17 December. He then changed his mind again, without telling White, and decided to attack on 15 December.

Buller ordered his artillery to bombard the hills along the Tugela River in an attempt to identify the exact Boer positions. But in spite of the heavy fire the Boers did not show themselves, and to some observers it seemed that the British guns were spraying shells harmlessly into an empty landscape. Dismissing such doubts, Buller instructed his brigade commanders to advance across a broad, open plain towards the river at dawn on 15 December.

But Buller did not really know where the Boers were. His brigade commanders were given vague hints that they were entrenched somewhere in the hills north of the river. However, as all the British maps were inaccurate, this was of little help. Major-General Fitzroy Hart was instructed to cross the river by a 'bridle drift' (a fording place), but as there were two, and both were wrongly marked on the maps, he can hardly be blamed for

At the Battle of Colenso in 1899 Colonel Long's placing of the British artillery too close to the Boer positions cost many lives.

not knowing where he was going. Major-General Hildyard was ordered to march to an 'iron bridge', but again there were two on the map, one for the road and one for the railway line, and there was no obvious reason to prefer one to the other. Major-General Lyttelton later complained that there had been 'no proper reconnoitring of the ground, no certain information as to any ford by which to cross the river, no proper artillery preparation, no satisfactory targets for the artillery'. Yet at the time no one raised any objections to the commander-in-chief's plan. Buller's reputation in the army was so secure that it is doubtful if anyone would have had the courage to question anything he said.

The next morning the British columns set out across the dusty veldt towards the river. The heat was already severe. At 5.30 am British naval guns began to bombard the hills on the far side of the river at a range of five thousand yards. However, the progress of the columns was about to be disrupted by one of the more extraordinary actions ever taken by a British artillery officer.

Colonel Charles Long was an officer with a great deal of military experience in India, and some very traditional attitudes. He believed strongly that 'the only way to smash those beggars is to rush in at 'em'. Supported by Brigadier Barton's infantry, Long moved twelve 15-pounder field guns, along with six naval guns, down the right side of a railway line. Buller had cautioned him not to risk advancing his field pieces at the start, and certainly not to move them any closer than two and a half miles from the river. Long, however, decided to advance the guns and close the range as quickly as possible.

At about 6 am, while still some three miles from the river, Long suddenly ordered his guns to gallop forward, leaving Barton's covering infantry far behind. The nonplussed Barton sent messengers forward asking Long to wait, but the latter went blithely on without taking any notice. When only a thousand yards from the river, and having left the naval guns six hundred yards behind and the infantry a further half mile, Long began to position his 15-pounders. He later admitted that he may have slightly misjudged the matter and gone too close. At once the Boers opened a withering rifle fire. Long proudly commented that his guns 'came into action in an excellent line', though whether this impressed the Boers is not recorded. None of the field-guns had shields to protect their gunners and

there was nowhere on the open plain for these men to take shelter. More than a thousand Boer rifles were concentrated on Long's unfortunate command. Long himself took a shrapnel wound in the liver, but seems otherwise to have gained nothing from his experience.

Buller, eating incessantly as was his habit during engagements, stopped chewing for a moment. His eyes widened in astonishment when he noticed Long's guns firing from so close to the river. He shouted to a staff officer, 'See what those guns are doing. They seem to be much too close. If they are under any fire causing severe loss, tell them to withdraw at once.' After an hour's firing, and having lost 12 men killed and 29 wounded, Long's command had exhausted their ammunition and were forced to scuttle towards the shelter of a gully, leaving the guns alone on the plain, in splendid isolation.

However, Long was not the only foolish commander at Colenso. Major-General Hart, commander of the Irish Brigade, was as traditional in his attitudes as Long and possibly even more lethally inept. It is difficult to decide whether Hart or Long was the braver man, or indeed the bigger liability to the British. Hart believed in keeping his troops 'well in hand', advancing shoulder to shoulder as if on an eighteenth-century parade ground. Against the massed rifles of the Boers this was suicidal.

Hart had begun the day with 30 minutes' drill for his brigade, as if this was an ordinary parade on the barrack square. He then ordered his battalions to march off in close order, himself riding at the head with an African guide. In broad daylight his men advanced towards the Boer positions across completely open ground in dense masses that provided the enemy with almost undreamed-of targets.

The officers on Hart's left flank could clearly see the Boer trenches packed with riflemen and sent three urgent messages telling him of this danger. But Hart would not be turned from his purpose and made it clear that he would ignore these Boers unless they attacked him in force. Ahead of him the Tugela – swollen from recent rains – flowed over a hundred yards wide and between fifteen and twenty feet deep – a formidable obstacle. Hart was following a clearly defined path towards the river when his African guide suddenly indicated that the ford they were looking for was some way to the right. He now ordered his brigade to veer right, following a huge bend in the river. All his maps showed that this was not the correct location of the ford and he should have sent scouts forward to find out where it really was, while sending a messenger back to inform Buller of the problem.

Even now disaster could have been avoided if he had spread his men in proper skirmishing order. Botha, meanwhile, found himself surrounding a British force on three sides without even having moved his position. Scarcely believing his good fortune, he opened a heavy fire on the dense mass of Hart's brigade from a range of a few hundred yards. As well as rifle fire the Boers also poured shells and grapeshot into the helpless ranks of the Irish. In the chaos the battalion commanders tried to deploy their men into more open formation, while Hart kept ordering them back into close order again. Any initiatives on the part of subordinate commanders were ruthlessly suppressed. Four thousand men were crowded into a loop a thousand yards wide. Within minutes everything was in confusion, with the battalions hopelessly mixed.

There were many examples of individual courage on the banks of the Tugela that day, but all of them were wasted. No one knew where the ford was and Hart was now lost, his guide having panicked and disappeared. Some men, fixing bayonets, rushed into the Tugela, fifteen feet deep at that point, and either drowned or swam across to the other bank, to be shot down there by the Boers. Hart showed great personal courage, walking calmly amongst the men and urging them onwards. 'If I give you a lead, if your General gives you a lead – will you come on?' he called, at which many jumped up and followed him.

From his observation point on Naval Gun Hill, Buller could see that Hart was in difficulties. 'Hart has got into a devil of a mess down there,' he told Lyttelton. 'Get him out of it as best you can.' Lyttelton moved two battalions up to the open end of the river loop and called on Hart to withdraw his men under the cover of their fire. This was done with great difficulty, Hart's brigade having suffered 532 dead and wounded in one of the most futile operations of the entire war in South Africa.

Meanwhile Buller had decided that the main attack would have to be postponed until Long's guns could be rescued. He promptly rode to the gully where the gunners were sheltering and took over command. Without a thought for what was happening elsewhere on the battlefield, the commander-in-chief concentrated his mind on the sole task of regaining the guns. With 8,000 men still not committed, Buller could think of nothing more than saving the guns.

Buller was a heavy man and no longer young, and the strain of commanding his first battle was obviously affecting his judgement. Far from acting coolly he went where he felt he could be of most use as a soldier, but where in fact he would be least use as a commander. With men falling all round him, his staff surgeon killed, and he himself badly bruised by a spent shell fragment, Buller's obsession with regaining the guns grew and grew. He called for volunteers to try to save the guns, but the ensuing heroism – which earned the admiration of the enemy commander Louis Botha and a total of seven Victoria Crosses – proved both costly and futile. Buller was not insensitive to the appalling butchery, and eventually he called off the attempt. The decision shocked Long, who considered that his guns had been 'deserted'. Buller's view was that he did not believe anyone could have got the guns out and lived. Perhaps if he waited until nightfall the guns might have been recovered with considerably fewer casualties.

Buller decided to abandon the battle at 11 am. He was physically and psychologically shattered. He later claimed that his men were exhausted, and that if the Boers had crossed the river his men would have had the worst of a 'rough and tumble; on the river bank. But there was never any question of the Boers leaving their strong defences to engage in any 'rough and tumble' with a numerically superior British army. The Boers were not 'rough and tumblers' or experts with the bayonets; they were riflemen, sharpshooters – and thinking men.

After the battle came the recriminations. Buller faced the shame of having lost ten guns, and not one of them actually captured by the enemy but simply towed away after the British had left. Total British casualties amounted to 1,139 men; it is doubtful if the Boers suffered more than forty. Buller attributed the disaster to rank bad luck. One of his generals viewed the battle differently: 'One of the most unfortunate battles in which a British army has ever been engaged and in none has there been a more deplorable tactical display.'

CHAPTER 2: FIRST COMMAND YOURSELF

The Cowardly

Although it does not feature much in regimental histories, cowardice is far from unknown on the battlefield. In fact, cowardice – the failure to overcome perfectly natural fear of death or mutilation – has been widespread among ordinary soldiers throughout military history. What is not so common is for cowards to reach such a high rank that their behaviour actually affects the outcome of a battle or campaign. Yet there have been examples of this, perhaps the most famous of which concerned Lord George Sackville at the battle of Minden in 1759.

'Milord Sackville . . .'

Sackville was commander of the British cavalry at Minden, but when he received orders from the commander of the Allied army, Prince Ferdinand of Brunswick, to charge the fleeing French and ensure a great victory, he sat stock still and did nothing. Three times

The Anglo-German victory at Minden in 1759 was achieved in spite of the efforts of Lord Sackville, who was frozen by inaction when ordered to charge the fleeing French.

the order was repeated by different aides, but each time Sackville claimed not to understand it. When his deputy, the Marquis of Granby, tried to respond to Ferdinand's order and lead out the second line of cavalry, Sackville stopped him and the French managed to escape. Relations between Ferdinand and Sackville had been bad for some time before the battle and it may be that Sackville was just being difficult. But King George II took a different view, court-martialling him on a charge of cowardice and cashiering him from the army with the heavy sentence that he should never again serve the king in any capacity whatsoever. Sackville was lucky not to be shot; Admiral Byng was executed for far less.

The flight of Gideon Pillow

During the American Civil War there was a surprisingly high number of acts of cowardice by commanding officers. Perhaps the worst case was during the siege of Fort Donelson in Tennessee in 1862. Besieged by Union forces led by Ulysses S. Grant, the Confederate commanders decided that they had no choice but to try to cut their way out and reach Nashville. In a snowstorm the Confederates poured out of the fort and drove back the surrounding Union troops. But having achieved this breakthrough, General Gideon Pillow suddenly lost his nerve and prevailed on his fellow generals, Floyd and Buckner, to call off the escape bid and return to the fort. Here the argument continued, with Pillow changing his mind again and arguing for another escape attempt, while his partners, who had wanted to escape earlier, now insisted that there was no alternative but to surrender. In fact, Pillow and Floyd made it clear that they had no intention of being taken by Grant. As politicians as well as generals they were afraid that they might be tried for treason and executed. What happened to the garrison of 15,000 men did not appear to concern them. Floyd, as overall commander, felt under the greatest pressure, so he simply abdicated command to his second, Pillow. Pillow was too wily a customer to fall for that trick, so he abdicated responsibility as well and passed command to Buckner, who was stuck with it. Floyd commandeered a steamboat and made his escape upstream to safety. Pillow had to make do with a small skiff, but as his slogan in 1861 had been 'liberty or death' he clearly felt justified in looking after his own liberty rather than risking death. The next day, Buckner surrendered the garrison unconditionally to Grant. He felt rather nettled about this as in 1854 he had once helped a down-and-out Grant get home by loaning him money. Such is the ingratitude of war.

Brock's bluff

During the War of 1812 between Britain and the United States the performance of the American general William Hull was a major cause of the collapse of the American invasion of Canada. Hull had been a good soldier in his youth, but at 58 felt unequal to the task given to him. Only presidential pressure made him accept the Governorship of Michigan and command of the left flank of the American attack on Canada. This involved a 200-mile march from Cincinnati to Detroit through wild Indian country against an enemy of unknown strength. Curiously, Hull asked his daughter and her two young children to share the perils of the march with him.

Hull collected 1,500 men of the Ohio Militia at Dayton, but immediately found that they thought his ideas on military discipline rather quaint, when he tried unsuccessfully to stop them 'tarring and feathering' one of their officers. The march proved arduous – there was no road and they were obliged to build it as they went – averaging only three miles a day as a result. In heavy rain and beset by blackflies and mosquitoes, the column gamely battled through the wilderness.

The problem for Hull was that no one had told him yet that war had actually been declared. Even when he reached the Maumee River, which joins Lake Erie, he still did not know for sure. This perhaps explains his next decision. Spying a schooner at anchor in the river, he decided to requisition the boat to transport the officers' baggage, the medical supplies, the tools and a trunk containing the army muster rolls – these contained the full details of the column and Hull's official orders and campaign details. As the boat sailed upriver to the lake, Hull expected to be able to recover the important cargo when it docked at Detroit. However, war had broken out and the British immediately seized the schooner as it passed Fort Malden on its way to Detroit. Why, one might ask, had Hull not been told about the outbreak of hostilities? The answer, it appears, was that the Secretary of State for War had sent the vital dispatch via the normal mail to the postmaster at Cleveland with the request 'please forward'. The news reached Hull the day after the schooner had sailed.

A week later Hull reached the Canadian border and was ferried across to a position north of Fort Malden. The British garrison at Malden was so weak that it was preparing to withdraw, but, as the Americans showed no aggressive intent, they stayed and were soon reinforced by a British naval vessel with eighteen guns. It was Hull's misfortune to have as an opponent the active and able General Isaac Brock, who immediately rushed more troops to Malden while the Americans looked on and did nothing. Rumours reached Hull, probably started by Brock, that 1,700 Indians were approaching and that 5,000 more were to follow. This seemed to break Hull's nerve, for he immediately withdrew to Detroit in American territory and retired to his room, where he remained for four days without seeing anyone. In the meantime, his troops occupied themselves as best they could. Attempts to reinforce Hull were twice ambushed by the British and Indians led by the Shawnee chief Tecumseh.

Across the river Brock was using psychological warfare to the full, dressing local villagers in red tunics to resemble British regulars and allowing a letter to fall into Hull's hands which said that there were now 5,000 Indians at Malden and that more would be superfluous. Hull's old eyes now saw visions of thousands of British regulars facing him as well as hordes of blood-crazed Indians. The presence of his daughter and grandchildren can hardly have made him feel more confident, 200 miles from his base and heavily outnumbered. The truth – which he could not see – was that Brock had a force approximately equal to his own.

Brock now felt that he had 'softened up' his opponent and asked for a parley. Backed by some six hundred Indians wearing war paint and ugly expressions, Brock tried to reason with Hull as 'between gentlemen'. He agreed with Hull that the Indians were rather unsavoury, but what could he do? Once their blood was up he could not, with the best will in the world, prevent them from doing the most terrible things. It would be best if Hull saw sense and surrendered. At first Hull refused and placed several cannon packed with grapeshot outside his fort, but as Brock's troops advanced towards him he abandoned them and called his men inside the stockade. Suddenly a British shell exploded inside the fort, causing several casualties, and without firing a shot in return Hull ordered the white flag to be raised. Brock could hardly believe his luck. To compound the ignominy a force of 350 men under two of Hull's most able commanders, whom he had sent south looking for help, returned to Detroit only to find that they had been included in the surrender and were immediately disarmed.

Hull, on his return from captivity, was court-martialled for 'cowardice' and sentenced to be shot. President Madison showed clemency to him on the grounds of his fine war record in the War of Independence. Perhaps Hull was more a dupe than a coward.

'Black Jack' Slade

Wellington's Peninsula generals included so many incompetents that the Iron Duke must have wondered sometimes who was the greater enemy, Napoleon or a British War Office that saddled him with men like 'Black Jack' Slade. Not only was Major-General Slade known to be inept but he was also a coward. Wellington in fact only kept him in Spain as the lesser of two evils, the alternative presented to him being a yet more egregious blunderer in the person of Sir Granby Calcraft. It was during the Corunna Campaign of 1809 that Slade had shown his true colours. Lord Paget reported that when Slade was ordered to charge with the 10th Hussars he kept stopping to adjust his stirrup leathers until the need for the charge had passed. On another occasion, when ordered into action he found it necessary to deliver a long lecture to his men which prevented his taking part in the fight. Sir John Moore had once noticed Slade carrying a message to the rear during a battle – a task that would ordinarily have been carried out by a junior aide. Only days later Slade brought Moore a report which he said he had been asked to deliver by one of his officers, Colonel Grant. Moore coldly inquired of General Slade whether he had now become Colonel Grant's aide-de-camp. Remarkably enough, Slade came to control the entire British cavalry in Spain in 1811 by seniority. 'He now let no possible opportunity for inaction pass him – pretending not to comprehend orders, which the events passing before him would have made comprehensible to a trumpeter – complaining that his hands were tied, and letting the opportunity slip.'

The Rash

'Every soldier actuated by the principle of cool and deliberate valour will always have the advantage over wild and precipitate courage.' These sensible words could serve as advice as much today as in 1814, when they were given by General Robert Rollo Gillespie to his troops at the siege of Kalanga fort in Nepal. Unfortunately, Gillespie did not practise what he preached and his own rashness brought disaster on himself and his troops.

In a pickle

Once described as 'the bravest man who ever wore a red jacket', General Gillespie was also a sadistic, drunken megalomaniac. By 1814, during his expedition to Nepal,

Gillespie was said to be prone to 'wild freakish behaviour'. Accusations about his private life had so embittered him that he was determined to wipe out any stains on his character by great military exploits, whatever the cost. Some believed that war had become a drug to him and that fighting was his only natural release.

Gillespie's small army was made up of 4,400 soldiers and about 5,000 camp followers and servants. Their baggage was considerable, one infantry major needing six camels to carry his personal liquor supply. Gillespie broke the force up into a number of detachments and ordered Colonel Sebright Mawby with 1,000 men to capture Kalanga fort, while he reconnoitred some positions to the northwest. Mawby found the fort to be immensely strong, although garrisoned by no more than 250 Nepalese soldiers. Considering his force too weak to besiege Kalanga, Mawby sent a message telling Gillespie that he needed more men and much heavier artillery. Gillespie replied that he would come himself and ordered the other parts of his army to march on Kalanga. The general was determined to take the fort in person.

Arriving at Kalanga, Gillespie now decided to attack the fort from all sides simultaneously. Dividing his force into four parts, Captain Fast was sent to attack from the northwest, Major Kelly from the north, Captain Campbell from the east, while the main force under Major Carpenter would assault from the south. It was absolutely crucial to co-ordinate these attacks. Now, Gillespie may have been a maniac, but was no fool. He realized the danger of sending his men off with native guides, who could easily get them lost. He therefore arranged for a signal of cannon fire to be made at ten o'clock, exactly two hours before the assault, to allow the separate columns to get into position for a mass attack at midday.

At dawn on the day of the attack he ordered his artillery to begin a bombardment of the fort to soften it up for the assault. But the guns failed badly and Gillespie's anger began to increase alarmingly. Embarrassed by the humiliating spectacle of British troops being defied by so paltry an enemy, he seems to have lost control of himself. His reputation for acts of courage beyond the capacity of other men convinced him that he could do anything. He now ordered his officers to prepare the attack for eight o'clock, a complete change from the instructions he had given to the supporting columns, now somewhere in the hills around Kalanga. Gillespie dismissed his officers' pleas for more time and ordered the cannon signal to be made. Not surprisingly, none of the support columns heard it, for the simple reason that they were not listening for it at that time.

Three companies of the 53rd Regiment had arrived at Kalanga the previous night after a march over appalling terrain. In view of their exhausted condition, Gillespie had ordered them to rest in the camp. But now he sent a message ordering them straight into action. Acting on the general's earlier order the regiment was relaxing and quite unprepared for action. Furthermore, the officer designated to lead them was not even in camp.

The situation now deteriorated rapidly as Gillespie's grip on reality slipped yet further. Major Carpenter ordered his infantry to advance to the attack, but so heavily loaded were they that they made slow progress. Sixty dismounted dragoons, placed at the front, rushed off towards the fort and were hacked to pieces by the Nepalese before the British infantry could intervene to help. To compound their problems the hut housing the scaling ladders burned down and their means of scaling the walls was lost. In the hills to the north and east, the commanders of the supporting columns heard confused firing, but, checking their watches, concluded that it was part of the 'softening up' process and was not a signal for them to join in the attack. Meanwhile, the wretched men of the 53rd Regiment at last appeared ready to join the attack, only for Gillespie to curse them as scoundrels. Their hatred of him was of long standing, and hours of marching over rubble and rocks had not improved their mood.

One of Gillespie's men now made an important discovery. On the northwestern wall of the fort was a wicket gate, which the defenders were covering with a cannon. Gillespie ordered a cannon to be brought up to fire on the gate. But the Nepalese fired first, blowing some of the 53rd infantrymen to pieces with grapeshot. The remaining redcoats now turned tail and fled. To Gillespie's fury the Nepalese scurried out from the fort and disabled his own cannon. Seeing what he took for cowardice in the men of the 53rd, he began screaming at them and then, pulling out his sword, rushed straight towards the fort as if to capture it himself. But within seconds he had been shot and died without saying another word. Ironically, it was at this very moment that the support columns burst out from the woods surrounding the fort, only to see the general's body being carried away and the main force retreating under Colonel Mawby's orders.

Gillespie was not mourned by his men. In fact even his officers were glad that he was dead. One of them wrote, 'It is the general opinion here that Gillespie's death has saved the army . . . As long as he lived, the Europeans would have remained unresisting to be killed.' Some wags were amused at the fact that the general's body was preserved in spirits, saying that he was 'a pickle when alive and a preserve when dead'. A cruel epitaph for a brave but rash commander.

Bloody Braggadocio Byron

At the Battle of Marston Moor in 1644 the rash behaviour of Lord Byron, one of the Royalist commanders, was to cost King Charles I dear. Commanding the right wing of the Royalist army, Byron had strict orders from Prince Rupert on what strategy to follow. Facing the parliamentary left flank of 5,000 men led by Oliver Cromwell, Byron's 2,600 men needed more than mere courage to succeed. Rupert had skilfully mixed musketeers with cavalry in Byron's front line and had emphasized that he must not quit his position. In front of him was marshy ground and a ditch, and before Cromwell's men could reach Byron they would have been slowed not only by the terrain but by musket and cannon fire from 1,500 musketeers under Colonel Napier. It was crucial therefore for Byron's cavalry to wait until Cromwell's troops had been disordered before making their attack.

Byron was not a patient man and waiting in position for much of the day stretched his nerves to breaking point. Surrounded by courageous but reckless young cavaliers, he must have faced countless questions as to why they had to submit to delay in the face of a persistent bombardment from the enemy. When Cromwell's cavalry erupted towards them, Byron's nerve broke and all the Prince's careful planning was forgotten. Ordering his cavalry forward, Byron charged pell-mell against Cromwell's troopers, scattering his own musketeers in the process. The ditch and the marsh which should have acted as a brake on Cromwell's progress instead slowed Byron's men and when the clash came they were swept away and broken, opening up the entire right wing of the Royalist army. Sweeping around the back of Rupert's army, Cromwell's cavalry next fell on General Goring's left wing cavalry, who had actually defeated their opponents. It was

A victorious Oliver Cromwell at Marston Moor in 1644. The recklessness of Lord Byron enabled the left wing of the Parliamentary forces to scatter Prince Rupert's cavalry, opening the way for a rout of the Royalist centre.

Cromwell's cavalry – and Byron's blunder – that won the battle for Parliament and lost the King the whole of the north of England – an expensive price for disobedience.

Where courage is not enough

While physical courage is rarely absent from military commanders, all too often the vital ingredient of moral courage is lacking. The decision to refuse a challenge – to stay out of the fighting and not to risk one's life – can be a hard one for a military commander to make. But the responsibility of command sometimes imposes limitations on how courageous a general can afford to be. Not that this has prevented generals throughout history from throwing away their lives, as if they were common troopers, without a thought for the welfare of the rest of their men or the outcome of the overall fight.

At Flodden in 1513 the Scottish commanders, true to their medieval chivalric beliefs, and apparently having missed out on the Renaissance revolution in generalship, all fought in the front ranks of their army. They sneered at the English generals for skulking at the back, particularly the elderly Earl of Surrey, who was crippled with rheumatism. The facts speak for themselves. Without proper direction the more numerous Scots were beaten by the English. But as most of the Scottish lords died along with their king, defeat turned into a rout, with no one left to organize an orderly retreat from the field.

At the battle of Vionville in 1870 during the Franco-Prussian War, French Marshal Bazaine, as famous for his personal courage as his fellow incompetent Redvers Buller, galloped up to the firing line, urging his men forward by his personal example, siting guns and leading counter-attacks. The fact that he was in overall command of the French

Army seems to have been forgotten. At Gravelotte-St. Privat (see p. 117) his inability to command rather than merely fight led to chaos in which for long periods none of his panicky aides were even able to find him. The French army fought without direction through the great battle, and one historian has written of Bazaine that his conduct 'can best be compared with that of a simple soldier who abandons his post in the face of the enemy'.

Gaston de Foix, bravest of commanders during a 'heroic age' of French warfare, threw his life away in a cavalry pursuit of a beaten enemy at Ravenna in 1512. A more pointless loss can scarcely be found. In all the above cases rashness was merely the outward sign of a more profound disease: too much courage and too little sense.

'You need not worry your head about the Boers fighting. I undertake to lead my regiment through South Africa from one end to the other, armed with only pickhandles.'

Colonel Tucker of the 80th Foot, South Africa, 1880

Such bravado and contempt for the enemy contributed to the disasters of the First Boer War in which the Boers inflicted defeat after defeat on the British regulars.

The Quarrelsome

One might suppose a military commander had enough enemies without fighting his fellow generals, but instances of quarrels – even violent ones – between commanding officers have been common. Their effects, while always harmful, have occasionally been decisive. Perhaps the most famous such disagreement involved Generals Pavel Rennenkampf and Aleksandr Samsonov at the battle of Tannenburg in 1914. These two Russian generals had hated each other since brawling openly on a railway platform at Mukden, during the Russo-Japanese War in 1904. Ten years later, during the campaign which culminated in the German victory over Samsonov's Second Russian Army at Tannenburg, German Intelligence was able to assure General Ludendorff that there was no likelihood that Rennenkampf's First Army would come to Samsonov's aid if he were threatened. As General Hoffman, the brains behind the German plan, is reported to have said, 'If the battle of Waterloo was won on the playing fields of Eton, the battle of Tannenburg was won on a railway platform at Mukden.'

The Latin Colonel

When relations between generals are so bad that a commander keeps his plans secret even from his own side, then disaster is certain. In 1709, when Charles XII of Sweden was campaigning in the Ukraine against Peter the Great of Russia, he was seriously wounded and had to surrender command to Field Marshal Count Karl Gustaf Rehnskold. Immediately jealousies began to surface among the senior commanders. In particular, Rehnskold could not tolerate General Lewenhaupt, the commander of the infantry. The two men could not exchange words without a violent row erupting.

Lewenhaupt was a complex man, whose academic background had earned him the title of 'The Latin Colonel' among his men. He suffered from deep depressions in which he saw life as a conspiracy aimed against him personally. He had no real friends and regarded his junior officers as rivals rather than companions in arms. Nevertheless, he was extraordinarily brave and, when left to himself, a capable leader.

At 58, Rehnskold was still a great cavalry commander. Only loyalty to Charles, his protégé, had kept him in the field years after most men would have retired. He was truly 'married to the army' and had not taken leave in nine years of campaigning, spending every winter in camp. It is possible that he was unaware of the physical and mental deterioration that must inevitably follow so prolonged a period of strain, poor food and little sleep. The irritability and nervousness resulting may well have contributed to his growing hostility towards Lewenhaupt. Thus were the fortunes of the Swedish army now entrusted to two men, both brave and able, but both seriously flawed. Under the pressure of leadership the irritable and choleric Rehnskold brought out the worst in the paranoiac Lewenhaupt. Without Charles's guiding hand this clash of personalities brought disaster to their country.

The Russian army of 40,000 was in an entrenched camp near Poltava and Rehnskold discussed with his bed-ridden king the best way to attack it. Only Colonel Siegroth, acting as Chief of Staff, was privy to the plan, which was kept secret even from the Swedish officers to prevent any chance of a leak. The King assumed that his Field Marshal would inform Lewenhaupt, who was to be second-in-command. But he was wrong; Rehnskold told Lewenhaupt nothing. All the Marshal's second-in-command knew was that his men were to advance at a certain time and place. The only infantryman present who knew of the plan was Siegroth, the 'colonel of the day', and he took the secret to the grave, killed as he was at the outset of the battle.

Ignorant of the commander's plan, Lewenhaupt moved his 2,400 infantry forward on his own initiative, directly towards the Russian camp containing some forty thousand men. Bypassing a ravine and under heavy fire, Lewenhaupt overran two Russian redoubts, without cavalry or artillery support, and was on the point of storming over the southern rampart of the Russian camp with sword and bayonet when a messenger from Rehnskold arrived to recall him. Lewenhaupt was furious – once again, he believed, his enemy was intent on thwarting him at the moment of his greatest triumph.

In fact, faced by overwhelming odds, Rehnskold had decided to call off the battle and ordered Lewenhaupt to form up in marching order for the retreat. Lewenhaupt, now at the head of the column, claimed he received an order from the rear, where both the King and Rehnskold were placed, to halt and turn towards the enemy. As he prepared to face the Russians a furious Rehnskold rode up and accused him of disloyalty and of attempting to sabotage the army. Lewenhaupt replied that the order must have come from someone and that Rehnskold was now trying to make him take the blame for his own mismanagement. Whatever the truth, the result of Lewenhaupt's turn was to cause the Russians to pour out of their camp. A battle now could not be avoided.

It was therefore left to Lewenhaupt with just 5,000 infantry to engage with the Russian infantry, 24,000 strong and supported by 70 cannon. Rarely can such a unbalanced contest have taken place. Nevertheless, Lewenhaupt was at his best in these situations, especially when he was marching away from Rehnskold. As he said of the Field Marshal,

Peter the Great defeats the Swedes at Poltava in 1709. A bed-ridden Charles XII could only look on
helplessly as communication failure – with its roots in the mutual loathing of the Swedish
commanders Rehnskold and Lewenhaupt – handed victory to the Russians.

'I was utterly dismayed to find that neither with nor without orders could I please him; my mind was filled with a burning resentment and I longed to die rather than remain any longer under such a command.'

With drums beating and in perfect order the blue-coated Swedes advanced into the face of the Russian cannons, which Peter's Scottish general, Bruce, had placed in front of the Russian position to bear directly on the approaching Swedes. Lewenhaupt's only chance was to strike a hard blow to one part of the Russian line and then attempt to roll it up, with cavalry support. Without firing a shot the Swedes marched on, never wavering, even though cannon-shot cut swathes through their line and volleys of musket fire poured out from the Russian infantry. Incredibly, the Swedes burst through the first Russian line at bayonet point, sending the green-coats reeling back in panic. They captured a cannon, swung it round and began to fire into the Russian flanks. Having achieved what seemed impossible, Lewenhaupt waited for the cavalry to sweep forward and enlarge his breakthrough by rolling up the wavering Russian line. But it was not to be. Rehnskold, instead of leading the cavalry in support of the infantry, as so often in the past, was struggling with the control of a defeated army and experiencing the agonies of supreme responsibility.

The moment of panic in the Russian lines passed and as Lewenhaupt looked to the left wing of his infantry he saw that they had been shattered by a torrent of cannon shot, losing half of their number before even reaching the enemy. The Swedish army was breaking in half, and as the Russian infantry moved forward the remnants of the blue-coats went down fighting or took to their heels. Lewenhaupt rode up and down trying desperately to rally his men, but the battle was lost now beyond recall. The enigmatic Lewenhaupt, part hero, part villain of the battle, rallied the remains of the Swedish army and escaped with the King, while Rehnskold was taken prisoner.

The wounded Swedish king was persuaded to cross the Dnieper River to escape the pursuing Russians, while Lewenhaupt tried to save what was left of the army. Apparently overwhelmed by the responsibility, he took to his bed in the King's tent suffering from diarrhoea. Physically and mentally exhausted, Lewenhaupt was no longer competent to command. News of the approach of Russian forces under Prince Menshikov caused him to panic and he exaggerated the numbers of the enemy from ten to thirty thousand, at a time when his own force of nearly fourteen thousand could have matched them. On 1 July 1709, in a state of abject depression, Lewenhaupt surrendered his army without striking a blow.

Why did Lewenhaupt do this? One can only suggest a combination of factors: ill health, a disturbed mental state and a failure of moral courage. Neither Lewenhaupt nor Rehnskold had the qualities necessary to take over supreme command from the injured king. Under extreme pressure, their defective personalities proved unequal to the challenge and brought disaster to their army.

Squabbles domestic and foreign

The animosity between Lord Lucan, the commander of the British cavalry in the Crimean War, and Lord Cardigan, commander of the Light Brigade, is well known, and contributed to the disastrous management of the cavalry at Balaclava (notably in the catastrophic 'Charge of the Light Brigade'; see p. 98). The two men often came close to trading blows in front of their own troops. Such disgraceful and demoralizing squabbles were not restricted to the British army. During Napoleon's Egyptian campaign Generals Reynier and Destaing actually fought a duel, in which the latter was killed. Napoleon's marshals seem to have been a quarrelsome bunch. At Auerstadt in 1806 Davout claimed that he was 'betrayed' by Bernadotte, who held back troops from supporting him. Bernadotte's childish display of petulance left Davout with 27,000 men facing 50,000 Prussians. Only the remarkable stubbornness of the French soldiers saved Napoleon from a heavy defeat. During the Austrian campaign of 1809 Lannes and Bessières indulged in one of the fiercest disputes. Bessières, claiming to be insulted, demanded satisfaction. 'This very moment if you like,' said Lannes, drawing his sword. Only the intervention of Marshal Masséna stopped an immediate duel to the death. Masséna, treating the two combatants like naughty boys, told them, 'You are in my camp and I shall certainly not give my troops the scandalous spectacle of two marshals drawing on each other in the presence of the enemy.'

Some quarrels have been less violent but more insidious. In 1914, at Russian headquarters in Galicia, the Russian commander Ivanov and his chief of staff Alexeyev were on such bad terms that they would not speak to each other. Two copies of each telegram had to be sent, one to each of them. When orders were issued they were frequently contradictory as both men issued them separately.

Relations between Field Marshal French, commanding the British Expeditionary Force in 1914, and General Lanrezac, commander of the French Fifth Army, were poor from the start and hampered Anglo-French co-operation in the difficult days before the battle of the Marne. The two men seemed to take an instant dislike to each other, Lanrezac regarding French as a complete idiot, and the Englishman thinking the Frenchman 'not a gentleman'. Although neither spoke the other's language they decided to get by without interpreters in the interests of secrecy. The outcome would have been hilarious if it had not been so serious. At one stage French pointed at a map of the River Meuse and asked Lanrezac if he thought the Germans would cross by the bridge at Huy – in fact the only traversable bridge and the one which the Germans were crossing at that moment anyway. French's pronunciation was so excruciatingly bad that Lanrezac asked one of the British aides what he had said. When informed of French's query, Lanrezac replied, 'Tell the Marshal, I think the Germans have come to the Meuse to fish.' Eventually the French commander-in-chief, Marshal Joffre, had to remove Lanrezac in the interests of Anglo-French co-operation.

Inkermann

At the battle of Inkermann in 1854 during the Crimean War disputes between the two Russian commanders completely undermined the Russian effort. The mutual loathing between General P. A. Dannenburg and Prince Menshikov was to play an important part in a bungled operation. Early indications were not propitious, a lack of maps meant that some of Prince Menshikov's commanders were unsure of their orders and the prince, determined to keep Dannenburg in the dark, allotted him as small a role in the battle as possible. But Dannenburg, a veteran of the Napoleonic Wars, was not the man to take this sort of treatment lying down. He responded by virtually taking over command of the battle and gave instructions to his fellow generals in an attempt to make clear what Menshikov had left obscure. Where Menshikov had told General Soimonov to begin his attack at 6 am and turn left after climbing the Kilen-balka ravine, Dannenburg told him to begin his attack at 5 am and to turn right instead. Soimonov's

The surprise Russian attack on the French and British armies at Inkermann in 1854 was undermined by disputes between the Russian commanders.

division was supposed to work in conjunction with General Pavlov's, but no one thought to tell Pavlov that Menshikov's plans had been changed. Pavlov therefore expected to begin his attack at 6 am. Three regiments of Soimonov's division made their difficult ascent to the edge of the ravine and were ready to attack at 5 am, but Pavlov was not there to attack with them.

Soimonov's men had completely surprised the British and easily occupied their positions, but before the general could send back orders for the bulk of the division under General Zhabokritski to advance both he and his second-in-command were killed by accurate rifle fire. Zhabokritski never got his orders, and when at last he did advance it was much too late.

Pavlov, meanwhile, with 20,000 men, had arrived at his expected crossing point on the River Chernaya only to find that Admiral Nakhimov's sailors had not carried out Menshikov's orders to build a bridge. In fact, Dannenburg had contradicted the prince by telling Pavlov to arrange for the bridge to be built himself. The result was that neither Nakhimov nor Pavlov had carried out the work, each assuming that the other was doing it. Before Pavlov's troops could get up to support Soimonov's men, they had been largely wiped out. When Pavlov's men at last reached the rim of the ravine, Colonel Popov, Menshikov's chief of staff, appeared on the scene and ordered them to move forward to support the remnants of Soimonov's men, only for Dannenburg to arrive and order them to ignore Popov and move to the right. Popov and Dannenburg then exchanged words, and the latter rode off, having got his own way as usual.

The Russians may have heavily outnumbered the British, but their generals were fighting a separate war of their own. With Dannenburg and Menshikov pulling in opposite directions, the feeble old General Gorchakov, with 22,000 men, found it all too dangerous a business to get involved in. He contented himself with some desultory long-range gunfire but otherwise took no part in the battle.

While it is common in British histories of the Crimean War for the quality of the British generalship to be called into question, at least the British generals were all fighting on the same side. It is difficult to imagine a battle more ineptly conducted than was Inkermann by the Russian commanders. Menshikov took no part in the battle, indeed saw nothing of it at all, instead he was playing nursemaid to the Tsar's three sons, who happened to be visiting the front. It was Dannenburg who finally ordered the retreat to save the army – he told Menshikov – from complete destruction. Naturally the prince tried to lay the blame on Dannenburg, but the old disciple of Kutozov – the great Russian commander in the Napoleonic Wars – was a survivor, and it was Menshikov who lost his command.

The Ambitious

The British campaign in Mesopotamia in 1914-16 was one of the most ill-judged of any in World War I. Yet its original aim, the securing of the oil refineries around the port of Basra, was strategically sound. In November 1914 India Command sent a division under Lieutenant-General A. Barrett to capture the Shatt-el-Arab peninsula and the port of Basra, and within days he had done so against feeble Turkish resistance. Moving up the River Tigris, Barrett captured the town of Kurna, thereby establishing British control on the river to a distance of 120 miles from the sea. The campaign was a complete success and the strategic aims had been achieved at minimal cost.

At this point things began to go badly wrong. Two ambitious officers in the Indian army, General Nixon, the army commander at Basra, and Major-General Charles Townshend, who replaced Barrett as commander of the Sixth Indian Division, decided that if the Turkish army was as weak as it had seemed so far there was nothing to prevent them advancing up the Tigris, perhaps even as far as Baghdad itself. This was a glittering prize indeed, second only in importance to Constantinople to the Turks. Its loss would undermine their morale at a time when they were enjoying the best of the struggle with the British at Gallipoli. In fact, both men were allowing their personal ambitions to interfere with their common sense as soldiers. Fuelled by what were impossible dreams, given the limited resources available and the adverse conditions, Nixon ordered Townshend to move a further hundred miles from Basra to take the town of Amarah. This was an act of disobedience by Nixon, who was exceeding British campaign direc- tives. It was also a signal to Townshend that he had official support for an act of selfish bravado which would have tragic consequences for the men under his command.

Throughout his career Townshend had been driven by the need to pursue position and rank, and he had shown himself willing to stop at nothing to gain advancement. He was a very able general – and he knew it – but he had the irritating habit of letting every- body else know it as well. Townshend was once called 'the only real general in the entire Indian Army; but not exactly a gentleman'. He had a habit of criticizing his superiors, and of exceeding orders at every opportunity when in action. His justification was that he was prepared to be judged by the results he achieved. Where one enemy trench was the target he would secure three. At Atbara in 1898 he had won the thanks of Kitchener himself.

As a student of military history, he liked to imagine himself as following in the foot- steps of the great captains of history. In his conquests of Mesopotamia he would be a new Belisarius, earning himself titles like 'Governor of Mesopotamia' or 'Townshend of Baghdad'. As the 'spoils of victory' from Amarah, he chose a gigantic Persian carpet from the customs house, so heavy it took twelve Arabs to carry it, and had it sent back to his house in England to await the conqueror's triumphant return. Inspired by Napoleonic visions, Townshend took not just Amarah but Nasaryeh as well. His victories convinced him that this was but the start of one of the great military campaigns of history. He moved on and took Kut, planning next to strike at Ctesiphon and from there Baghdad. Rampant egomania had taken control.

Townshend was a brilliant general, but his victories were severely reducing the strength of his division. What others were calling 'Townshend's Regatta', because of the apparent ease with which he was beating the Turks, was in fact costing his men dear. At Ctesiphon he again beat a superior Turkish force, but now his numbers were so reduced that he was forced to withdraw to Kut, which he decided to fortify in order to withstand a siege.

As a young captain in Kashmir in 1895, Townshend had won renown by his defence of the small fort of Chitral against local tribesmen. Even though Major Aylmer VC had led the force which relieved Chitral, it had been Townshend who had been promoted, given the Order of the Bath, granted an audience with the Prince of Wales and invited to dine with Queen Victoria. If so much fame had come to a young captain for defending a minor Indian fort for 47 days, how much more might the defender of Kut expect?

During the siege that followed, Townshend's personality showed signs of great instability. He was so concerned with his own personal situation that he became indifferent to the suffering of his men or the success of his mission. In order to speed up the relief of Kut, he bombarded the authorities at Basra with reports of food and ammunition shortages, claiming that he was only hanging on by his finger tips. As fate would have it the man responsible for rescuing him was the same man who had saved him at Chitral, now Lieutenant-General Fenton Aylmer VC. Townshend's treatment of Aylmer was deplorable. His hysterical letters from Kut obliged Aylmer to make repeated attempts to reach the town, against heavy Turkish resistance. Each time Aylmer hoped that Townshend would make a sortie from Kut, or try to cut his way clear, but he did neither, simply staying put and waiting to be rescued. Aylmer's casualties were so heavy that after three failures he was relieved of his command and replaced by Gorringe, whose recent promotion to Lieutenant-General so mortified Townshend that he broke down and wept on the shoulder of a junior staff officer. On other occasions he suggested that he should personally escape and abandon his troops to their fate, and wrote ingratiating and possibly treasonable letters to the enemy commanders.

After a siege of 147 days – the longest in British military history – Townshend surrendered Kut and allowed his 13,000 troops to be led off on a death march into slavery from which 70 per cent never returned. He, in the meantime, was accorded the greatest honours by the Turks, living in a cliff-top villa on the fashionable island of Halki, in the Sea of Marmora, where he dined and hunted with Turkish dignitaries. His pursuit of a chimera, an Arabian Night's dream, had led to a disgraceful defeat for the British army. In terms of casualties, Townshend's ambitions had cost his country nearly 40,000 lives. Rather than the court martial that he deserved, on his return to Britain he was knighted, though he grumbled that he had merited a peerage. He later became a member of parliament.

A Greek tragedy

Visconti Prasca, an ambitious and scheming officer, was the real author of the disastrous Italian invasion of Greece in 1940. He was vastly unpopular with his fellow generals because he enjoyed close relations with Mussolini and they correctly suspected that the Greek operation was designed to further his personal career. Like Townshend, Prasca was a dreamer, who saw war against the Greeks as nothing more than a victorious progress against minimal opposition, which would culminate in a triumphal entry into Athens as a marshal of Italy. He inspired Mussolini with his own enthusiasm for a war against Greece which would be nothing more than a series of 'rounding-up operations against the Greek force'. Prasca used the hollow language of Fascism, speaking of 'liquidating' and 'shattering' the Greeks with his 'iron will'. In reality, as the more balanced Italian generals understood, none of this was possible.

Marshal Badoglio, for one, knew that Prasca was talking dangerous nonsense. Prasca's plans allowed for too few divisions, but when Badoglio suggested increasing the number of troops the cunning Prasca was able to convince Mussolini that he did not need them. In fact, as a junior Lieutenant-General he was only allowed to command a maximum of five divisions. If any more were employed, he would lose the command to a senior general. But if Prasca succeeded with his small force he could gain promotion over the heads of many of his rivals, and with the backing of Il Duce he felt confident that he could keep the command whatever his rivals tried to do. In October 1940 Mussolini told him that he had 'opposed all attempts to take your command away from you on the eve of the operation'.

So Prasca kept his command and condemned Italy to a truly disastrous campaign, riddled with corruption and muddle. Everywhere there were signs of haste and inadequate preparation. To achieve his ambitions Prasca had told a tissue of lies. His troops were too few and too ill-trained, and the Greeks were a far more powerful enemy than he had been prepared to admit. The result was that the Greeks effectively repelled the Italian invasion. It was only when the Germans intervened that Greek resistance crumbled.

THE ATHENIAN EXPEDITION TO SYRACUSE (415-413 BC)

The expedition against Syracuse during the Peloponnesian War was the greatest naval operation ever seen in the Greek world. Yet it was an unwarranted strategic gamble on the part of Athens, based on an inadequate knowledge of the size and geography of Sicily, and could only have succeeded if the Athenians could rely on substantial military aid and money from Sicilian cities hostile to Syracuse. If such help was not forthcoming, and if Syracuse gained help from Athens' enemy Sparta, then the expedition would be a failure. This much could have been predicted by the careful study of intelligence reports. What could not have been anticipated was the treachery of Alcibiades – the most brilliant Athenian commander – which robbed Athens of her chance of victory, and the superstitious weakness of another general – Nicias – which was to turn defeat into disaster.

Athens had already stretched its military resources to their limits in fifteen years of warfare, and for this – its greatest expedition – needed to hire large numbers of mercenaries to support its own hoplites. The Athenians were fortunate in having such leaders as Alcibiades and Lymarchus, but considerably less so in the limited support offered by their allies in Italy and Sicily. This factor should have been enough to persuade Alcibiades to cut his losses and sail home. Instead he reached a decision which was to bring ruin on the expedition. Alcibiades believed that it would be held as a disgrace if so great a force as Athens

had raised went home without achieving anything. He therefore recommended a cruise along the coast of Sicily, calling on city after city to support Athens. He gained some success, for example acquiring a naval base at Catana, but more significantly his actions gave the Syracusans warning that the Athenians were going to attack and gave them time to prepare their defences and contact their allies. A major blow to the Athenian cause was struck when Alcibiades was called home to answer charges of impiety – he was accused of knocking the faces and phalluses off religious statues – and was replaced by the less forceful and far less brilliant Nicias. Unwilling to face trial – and possible execution – in Athens, Alcibiades escaped to Sparta and was instrumental in persuading the Spartans to send help to the Syracusans in their struggle against Athens.

Throughout 414 BC the Athenians blockaded Syracuse, certain that victory was imminent. However, by settling for a prolonged siege the numerically superior Athenians were allowing the Syracusans to dictate the conduct of the war. This was brought home to them in crushing fashion when a Spartan relief expedition under the able Gylippus reached Syracuse. How the Athenians must have regretted the time they had wasted. Now the struggle became one of survival. Gylippus immediately completed the counter walls that the Syracusans had been building to prevent Nicias ringing the city with siege walls. The addition of Spartan and Corinthian ships to the Syracusan fleet meant that the Athenians were in danger of losing their supremacy at sea. To compound Athenian woes, Lymarchus was killed in a skirmish and Nicias was left in overall command. Even the arrival of reinforcements under Demosthenes did not relieve the depression in the Athenian camp. In order to enforce the siege the

An outstanding if occasionally wayward commander, Alcibiades planned and jointly led the Athenian expedition to Syracuse 415–413 BC. Called back to Athens to face charges of sacrilege, he defected to the Spartans, supplying them with valuable military advice that contributed to a calamitous Athenian defeat.

Athenians had to camp on low-lying, swampy ground, where they were prone to fever and sickness. Nicias himself was struck down by fever and became gloomy and morose.

By the summer of 413 BC the Athenians were clearly in trouble. Nicias was consigned to his bed by illness and the command passed to Demosthenes, who had only recently arrived on the island. He soon saw that there was now little chance of success and that the only option was to raise the siege and retreat while they still could. But the sickly Nicias persuaded him to delay the evacuation on the grounds that he believed that some of the Syracusans, if 'bought' with Athenian money, would force the Spartans to surrender the city to him. But this was wishful thinking and another month passed, during which more Spartan reinforcements arrived on the island. Demosthenes now pressed Nicias for an immediate withdrawal. Nicias, we are told by the historian Thucydides, was a great believer in omens and superstitions and when, during the preparations to evacuate the Athenian army, there was an eclipse of the full moon, he let his troops convince him that it portended disaster. According to Nicias there was no option now but to honour the fates and wait the 'thrice nine days prescribed by the soothsayers'.

This was a catastrophic decision and was to have remarkable consequences for the whole Greek world. While the Athenians waited at Syracuse for the soothsayers to be satisfied, the Spartan general Gylippus blocked the mouth of the harbour with a line of triremes and merchant ships chained together, cutting off the Athenians' escape route by sea. When the appointed day came for them to move, their only option was to try to escape into the interior of the island in two divisions, led by Nicias and Demosthenes. But the Syracusans and their Spartan allies were waiting for them and wiped out both parts of the Athenian army, killing their commanders and taking 7,000 prisoners, who were set to work in the island quarries. Of the Athenian force of 50,000 sent to Syracuse, virtually none returned. The episode marked the end of Athens as a great power. How different things might have been but for an eclipse of the moon and, in the words of Thucydides, a commander 'somewhat over-addicted to divination'.

THE BATTLE OF HATTIN (1187)

The battle of Hattin in 1187 was one of the most important battles of the Middle Ages. It led to the fall of Jerusalem to the Muslim leader, Saladin, and the collapse of the Crusader kingdoms in the Levant. Yet it was the decision of one man, Guy of Lusignan, which committed the entire military strength of the kingdom of Jerusalem to a campaign in a waterless desert. Flying in the face of reason and good sense Guy took the one decision that was certain to spell doom to the Crusaders.

In the years leading up to 1187 the kingdom of Jerusalem was split into factions. On one side were the native lords, descendants of the original members of the First Crusade, such as Raymond of Tripoli and Balian of Ibelin. Against them stood the 'new men', who had entered the kingdom more recently, such as Reynald of Kerak and Guy of Lusignan. And between these two groups were the powerful military orders, the Hospitallers and the Templars. The Grand Master of the latter order, Gérard of Ridefort, hated Raymond of Tripoli, the pre-eminent noble of the realm, and was willing to help the incompetent Guy of Lusignan to seize the throne on the death of the child-king Baldwin V in 1186. These political squabbles rent the kingdom at precisely the time that the Muslims of Egypt, Syria and Iraq had united under a single leader – Saladin – in a holy war to retake the city of Jerusalem. Saladin knew that he could not hope to beat the Crusaders if they stayed within their powerful castles of Acre, Kerak and Tyre. Only by drawing them out into a pitched battle could he hope to defeat them. Afterwards, once their garrisons were gone, he would be able to take the castles easily.

Saladin hoped to exploit the splits in the Christian ranks. He knew King Guy was a fool but that his advisers, notably Reynald of Kerak and Raymond of Tripoli, were powerful warriors and expert desert fighters. How could he be certain that Guy would advance to meet him in open battle?

In 1187 Saladin invaded the kingdom of Jerusalem and laid siege to the city of Tiberias on the edge of the Sea of Galilee. He knew that Guy had already stripped the castle garrisons throughout the country to assemble the royal army at Saffuriya. He reasoned that as Raymond of Tripoli was with the royal army and Raymond's wife was unprotected in Tiberias, King Guy would have no option but to advance to the aid of the lady and her city. When a messenger rode from Tiberias to Saffuriya calling for King Guy's help, Saladin made certain that he was not molested. Meanwhile he assembled the main strength of his armies on the high ground on either side of the waterless plateau which separated Saffuriya from Tiberias. There he would wait until his prey walked into his trap.

On hearing of the plight of the 'Lady of Tiberias', Guy called a war council and asked his senior commanders for their advice. Raymond of Tripoli spoke first and to everyone's surprise argued strongly against a rescue bid. He said that Saladin was planning an ambush and that if the royal army advanced into the waterless desert the whole kingdom could be lost as the result of one battle. He knew that Saladin, as a man of honour, would not harm his wife, and that Tiberias could

be recaptured once Saladin was forced to withdraw through lack of supplies. Angrily, Gérard of Ridefort, the Grand Master of the Templars, condemned Raymond as a coward, saying that it was a disgrace for the army to sit doing nothing while the Saracens besieged a Christian city. But Guy, and the rest of the lords, had been convinced by Raymond's arguments and the decision was taken to resist the temptation to rush to the lady's defence.

That night Gérard returned to Guy's tent as the king was going to bed. Guy was clearly frightened of the Templar, who had been instrumental in helping him to the throne, and under pressure reversed his decision not to march. Gérard had threatened that if he did not do so the military orders would withdraw from the army. The rest of the camp was shocked to hear that the king had changed his mind. Most knew that a march to Tiberias in the heat of midsummer was certain to end in disaster.

The Crusader army, probably 15,000 strong, with 1,500 knights, set out from Saffuriya on the fifteen-mile journey to Tiberias just before sunrise on 3 July 1187. They were leaving the well-watered gardens of Saffuriya for the sunbaked plateau ahead where they knew the Saracens would be waiting for them. The vanguard was led by Raymond of Tripoli, with Balian of Ibelin commanding the rear and the king with his lords in the centre. At first the Saracens confined themselves to hit-and-run attacks. The knights in the centre of the Crusader column were forced to move only at the speed of the footsoldiers, who formed a protective sleeve all around them and who – with their mail shirts and leather gambesons – shielded their horses from the Saracen arrows. But from the heat of the sun there was no protection.

The column had no water-carts, since these would have slowed it down even further, and instead the men carried their own water bottles. Before the sun had reached its height most had used up all their precious supply and had no prospect of more water until they reached the Sea of Galilee. With the column well and truly committed to the march, and retreat now as hard as going forward, Saladin launched the wings of his army, which charged down from the hills and began to harass the Crusader column. The Christian crossbowmen kept the Saracens at a distance, but, in order to fight, the column had to slow almost to a standstill. Meanwhile, above their heads the sun was taking its toll on the heavily armoured Crusader soldiers.

By 10 am the Crusaders had been marching for nearly six hours and were exhausted. A northward diversion could have brought them to a spring near Mount Turan, but King Guy ordered his men to march on and the chance of water was missed. In the vanguard Raymond was convinced that the long, straggling column would die of thirst long before it reached Tiberias and suggested to the King that they change direction and head for the springs at Hattin, which lay in rising ground to the north. This course of action would mean abandoning Tiberias for the moment, but it might save the column. The king agreed and the whole army now turned in a northeasterly direction near Meskenah. In the confusion of the turn some knights broke away from their infantry escort and tried to quicken their pace towards the water. Sensing that his prey might elude him, Saladin launched a force of cavalry under his nephew Taqi al-Din to ride around the column and block its path towards Hattin. Raymond realized that unless they could break through Taqi's men the whole column was doomed and prepared to charge the Saracens with his knights. Before he could do so a desperate and – as it turned out – fatal message reached him from King Guy, ordering a halt and the setting up of camp. The exhausted Crusaders could go no further that day. Without water Raymond knew that this was a sentence of death and told the king, 'Alas! alas! Lord God, the war is over. We are betrayed to death and the land is lost.' But Guy was adamant and the camp was set up.

With his enemy trapped on the plateau, far from any water, Saladin's troops now surrounded the Crusader camp, harassing the Christians all night with shouts of triumph and a cacophony of noise. In the light of their fires the crusaders could see the Saracens deliberately taunting them with offers of water, only to tip it in the sand as they held out their hands. The horror of their situation was exacerbated by scorpions and poisonous spiders which crept into their armour.

At first light the Crusaders began their march towards the springs at Hattin. Hundreds of horses had died from thirst or from the Saracen arrows and many knights marched with the infantry. Morale within the Christian army was very low, and some knights deserted to the enemy. The fighting slowed the column's momentum until it was again halted. Suddenly the infantry gave way and with a great moan hundreds of them abandoned the column and rushed up the rocky slopes towards where they thought water could be found. In desperation, Guy again pitched his tent and tried to rally his troops around him. To add to the suffering of the Crusaders the Muslims now lighted brush fires, which blew smoke constantly into their faces and increased their desperate craving for water.

Sensing that the day was lost, Raymond massed his mounted knights around him and charged the Saracens guarding the approach to the village of Hattin. The knights of Tripoli and Sidon cut through the enemy and made their escape into the hills, but for King Guy and the mass of the army there was no escape. Without infantry to hold back the Saracens with their crossbows, Guy's knights were almost helpless. Taqi al-Din charged at the royal tent and in the fighting captured the True Cross from the Bishop of Lydda. With the loss of this holy relic, the symbol of God's support in the fight, Christian morale collapsed. So exhausted was Guy that he simply sat on the ground and awaited capture. With him were taken most of the lords of the kingdom, except for Raymond of Tripoli and Balian of Ibelin, who had escaped.

Saladin treated his noble prisoners well, though the common soldiers were all sold into slavery. So complete was the Christian defeat that every city and castle within the kingdom – save Tyre – fell to Saladin in a single campaign. Against the advice of more experienced soldiers Guy had risked everything on a single battle. He had led his army into a trap, even though he had been warned against it. He fought Saladin in conditions that suited his enemy rather than himself, and had twice halted his column on a waterless plain when he must have known that to do so would be fatal. It would be an understatement to call Guy a fool: brave, certainly, but too incompetent to command a Crusader army.

THE BATTLE OF THE CRATER
(1864)

It is difficult not to like that most personable of incompetent commanders, Ambrose Burnside – a bewhiskered buffoon who gave his name to 'sideburns'. While he may have committed his greatest blunders two years earlier at Fredericksburg (see p. 116), he saved something rather special for the siege of Petersburg in 1864. As the Union forces under Ulysses S. Grant pressed ever closer to the Confederate capital at Richmond, the Union IX Corps under Burnside were positioned at a crucial point opposite an enemy strongpoint named Elliott's Salient. Henry Pleasants, the colonel of one of the regiments under Burnside's command – the 48th Pennsylvania – had been a civil engineer before the war and believed he could tunnel under the Confederate position and blow it sky high.

Burnside was impressed by the plan, although the Union army engineers tried to ridicule it, saying that Pleasants could never burrow 500 feet without cave-ins and ventilation problems. Burnside managed to persuade Grant to allow Pleasants to put the plan to the test, but the engineer met obstruction from the army professionals, who refused to lend his men the necessary picks, spades, wheelbarrows and timber to shore up the shaft or even a theodolite for Pleasants to take his readings. Undeterred, Pleasants overcame every setback and by 17 July 1864 was able to tell Burnside that the tunnel was ready. Pleasants now sat back to see what use his Corps commander would make of his masterpiece. He was to be very disappointed.

Grant was quite taken by Pleasants' idea and thought it might lead to the decisive breakthrough he had been waiting for. He promised to back Burnside with extra troops and masses of artillery. While Pleasants packed the mine shaft under the Confederate position with four tons of high explosive, Burnside planned the infantry attack which would follow the explosion of the mine. Of his four divisions, his immediate choice to lead the attack was an all-black one commanded by General Ferrero. The black soldiers were proud to be chosen and eager to attack. But Burnside's choice

General Ambrose Burnside was, by his own admission, not competent to command the Union forces in the American Civil War. His appointment was an error of judgement on the part of President Lincoln. The blunders that followed at Fredericksburg and Petersburg were Burnside's own.

was vetoed by Grant, on the grounds that the mission was dangerous and it might look as if they were prepared to throw away black lives. Burnside took Grant's decision badly and seems to have lost interest in the whole affair, eventually selecting from his other three divisions by pulling straws. His now haphazard choice fell on General Ledlie, an unreliable commander, suspected of being both a coward and a drunkard. Even Burnside had described Ledlie's men as 'gun-shy' and 'worthless', so why he let them lead the attack is difficult to understand.

Ledlie's troops assembled in the forward trenches in darkness, awaiting the firing of the mine at dawn. It seems that no one had told Ferrero's black troops that they were no longer leading the attack and so they also assembled. Burnside, on the other hand, took cover half a mile from the front line and played little part in what followed. The fuse took some time to catch and the explosive was ignited some ninety minutes late, but when it did its effect was shattering. So enormous was the blast that many of Burnside's own men fled. Once Grant's 80 heavy guns began pumping shells into the smoking ruins of the Confederate position all that was needed was for Burnside's infantry to sweep through the crater left by the explosion and on to take Petersburg.

It was now that a major problem was discovered. Since no one had given orders for the parapets to be lowered Ledlie's men found themselves trapped at the bottom of eight-foot-deep trenches, unable to climb out in battle formation. Some managed to improvise ladders with their bayonets, while others slowly gathered piles of sandbags. As a result of this delay the anticipated forward rush of Ledlie's division became instead a mere dribble. When the Union troops reached the crater they found sixty yards of Confederate line destroyed, replaced by a hole thirty feet deep. Amazed at the sight, Ledlie's men climbed into the hole, some helping to rescue buried Confederate soldiers, others digging trenches as if their task was to fortify the crater. Where were the commanders who should have been driving them on to victory? The divisional commander, General Ledlie, was hiding in a bombproof shelter a quarter of a mile back, drinking rum and refusing to give any orders. Even further back, Corps commander Burnside simply kept ordering more troops forward without bothering to leave his shelter.

Two more divisions passed down the Union trenches, scaling the sandbag or bayonet ladders one by one, then crossing 'no-man's land' and climbing into the crater. But once there, they advanced no further and soon the crater was packed with thousands of Union troops. The Confederates meanwhile were recovering and had begun to fire on the dense, heaving mass of men in the middle of their position. Burnside's army commander Meade lost his temper and began shouting at him in 'an unofficer-like and ungentlemanly' way. Possibly the greatest chance of ending the war early was lost as Meade and Burnside abused each other, and Ledlie drank himself under the table.

Ferrero's black division, unaware that the plan was in chaos, charged towards the Confederate trenches and captured some of them. But when runners were sent back to request orders as to what to do next, General Ferrero, who had stopped off at Ledlie's shelter and joined his drinks party, told them just to capture whatever came next. In fact, the Confederates were furious to see black soldiers in their trenches and showed no mercy: 'Take the white man, kill the nigger' was their motto. Soon Ferrero's division was fleeing back to the Union lines in tatters, with fifty per cent casualties.

Colonel Pleasants, meanwhile, forgetting army etiquette, was yelling at Burnside that his commanders were nothing but a bunch of cowards. Grant tended to agree, viewing the fiasco as 'the saddest affair I have witnessed in the war'. The problem remained of how to save the 10,000 men trapped in the crater. The decision was taken out of Burnside's hands when the Confederates massed troops for a counter-attack and emptied the crater, killing hundreds and taking thousands more prisoner.

Grant had seen enough: IX Corps must have a new commander. Burnside was granted an extended period of leave and retired from the service, later to enter politics. Ledlie was rightly court-martialled for drunkenness and cowardice, but Ferrero, curiously, escaped punishment. Thus one of the most brilliantly conceived operations of the Civil War ended in bitterness and recriminations.

THE BATTLE OF CUSTOZZA (1866)

That part of the Austro-Prussian War of 1866 played out between lumbering armies of Germans in Bohemia has perhaps a certain Wagnerian flavour. The war's southern front – boasting a cast of Italian generals with mellifluous names performing comic antics in the idyllic surroundings of Lake Garda – undoubtedly has about it something of *opera buffa*. As in a Shakespearian comedy, General Lamarmora and King Victor Emmanuel II appear in numerous guises, confuse everyone including themselves, and turn serious matters into farce.

The aim of the Italian army of 175,000 was to capture the province of Venetia, precariously held by 75,000 Austrians under the command of the Archduke Albert. On paper the task was not a difficult one, but the Italians contrived to create problems where none existed. In the first place, there was confusion as to who was in command. The main army of 100,000, which was to cross the River Mincio, was notionally commmanded by General Lamarmora, but with the king, Victor

Emmanuel, accompanying the army, Lamarmora found it impossible to exercise proper control. The second army of 75,000, which was to cross the River Po, was commanded by General Cialdini. But was Cialdini subordinate to Lamarmora? And was Lamarmora subordinate to Victor Emmanuel? And did Victor Emmanuel know what he was talking about anyway? If these questions had not been answered to the satisfaction of the main participants, one can imagine the difficulties that their subordinates faced. The Italians lacked co-ordination from the outset and this was to have grave consequences. Matters were worsened by the lack of telegraphic links between the two armies, so that for hours at a time they were completely ignorant of each other's movements.

If the Italians were confused about what they were going to do once they had begun their attack, they had given absolutely no thought as to what the Austrians might do. The king and Lamarmora seemed to regard the Archduke's army as something static, like a fortress or town, around which they could manoeuvre as they wished. The truth was not so kind. The Archduke was an able commander who had thoroughly studied the terrain and had forged a strong and coherent strategy. As the Italian troops advanced, without scouts and with the cavalry regiments eccentrically placed at the rear, the Austrians launched a surprise attack near Custozza. Lamarmora was caught completely unawares, his staff officers not even having crossed the river yet. Most of his twelve divisions were not deployed for action, and the Austrian flank attack threw the entire Italian army into disorder. Lamarmora now lost his head and instead of trying to improvise a headquarters galloped around the battlefield, seeking out individual commanders and giving merely local orders. Extraordinary to relate, at one point he rode twenty miles to consult with his staff on the other side of the river about arrangements for a retreat. All this while the fighting was going on, undirected, on the Italian side.

To make matters worse, the king set up his own HQ and began issuing orders of his own, frequently contradicting what Lamarmora had said. Some Italian historians have described the king as being in the front line throughout the day, but the truth was that having done what damage he could, he left the battlefield and spent most of the day twenty miles away on the other side of the river.

Italians have described the battle as 'one of the most bloody battles in all modern history', but this is fanciful. Italian losses were some 3,500 with 725 dead; the Austrians rather higher. But through sheer blind panic the Italian generals managed to turn defeat into disaster. The Austrians, satisfied at having checked the Italian advance, expected the fighting to continue the next day. But Italian morale, so inflated by false expectation before the battle, had now slumped and the generals were quite incapable of restoring it. Lamarmora and the king saw only confusion around them and lack of experience convinced them that they had suffered a far worse defeat than was in fact the case. Fearing an Austrian pursuit, they could think of nothing but an immediate retreat and destroyed the bridges over the Mincio.

Cialdini, awaiting news from Lamarmora about his crossing of the Mincio, received a series of telegrams from the king, each bearing gloomier tidings than the last. One spoke of the army suffering 'immense losses' and a later one mentioned 'an irreparable disaster' and the need to fall back to 'cover the capital' – Turin. If the king cannot be blamed for the early mistakes at Custozza, he certainly can for the way he turned an army which had suffered a reverse into a beaten rabble, eager only to escape. Cialdini felt he had no option but to abandon plans to cross the Po and retreat. After all, the Austrians might now change fronts and attack him. Retreat was the only safe answer. Thus were two Italian armies defeated by a force less than half as strong. Cialdini blamed the disaster on the king, though protocol prevented him saying so in public. In later years he referred to his monarch as someone who 'understands absolutely nothing at all of what is going on . . .' and as 'wholly ignorant and incompetent.' Another commentator said of Victor Emmanuel that he was, 'an incorrigible braggart . . . who thought that there was nothing more to the art of war than courage, of which he had more than anyone . . . He conceived himself to be a great captain.' While one must blame Victor Emmanuel for interfering and sending damaging telegrams, one must also call into question the professionalism of Lamarmora and Cialdini, both of whom panicked when confronted by a single setback. They lost their nerve and, more than the ordinary soldiers who fought at Custozza, were the ones who brought ignominious defeat on the Italian army.

CHAPTER 3: PAINTING THE WRONG PICTURE

The Impetuous

In warfare courage has too often been regarded as an acceptable alternative to careful planning and forethought. French military thinking immediately before 1914, influenced by the views of Colonel Grandmaison, saw in the courage of the French soldier a way of overcoming German advantages in numbers, armaments and logistics. War memorials all over France bear witness to the falsity of this argument. But to find the most spectacular examples of impetuosity it is best to turn from the twentieth century to the Middle Ages.

'Victory shall be gained by the breasts of our infantrymen.'

Colonel Pedoya, August 1914

This horrifying quote – horrifying because it was literally true – reflected the current French belief in all-out offensives. This attitude contributed to the massive French casualties in August 1914.

Robert of Artois

The 7th Crusade under Louis IX (St Louis) of France offers an example of a commander 'painting the wrong picture' – acting from a completely incorrect, or in this case, inadequate assessment of the military situation. Having captured the coastal city of Damietta, at the mouth of the Nile, Louis led the Crusading army upriver towards the town of El Mansura, where the Egyptian army was camped, defending a canal which Louis needed to cross. The Christian engineers tried to build a causeway, but even as they did so the Muslims destroyed it with catapults and siege engines. For two months the Crusaders struggled to find a way of crossing the canal until they discovered a ford some four miles away, which would be safe for their cavalry to use. Once across, the knights could attack the Egyptian camp and destroy the siege engines. Louis knew that he was taking a calculated risk by using an unknown ford, but he had decided that unless the causeway could be completed the Crusade would have to be abandoned.

The crossing of the ford required immense discipline. No rider was to leave his position in the column, nor was the vanguard to set off on its own until the whole force had crossed. However, Louis should have known better than to place his brother, Robert of Artois, at the head of the vanguard. Discipline was not one of Robert's strongpoints – nor was obedience. Once safely across the canal, Robert sighted a group of 300 Arab

A scene from the seventh crusade under Louis IX of France, showing Crusaders cut off by Saracens in a tower. Impetuosity and lack of regard for his Muslim adversaries led Robert of Artois to disaster in the narrow streets of El Mansura.

horsemen and decided to give chase. Some of his fellow knights told him to remember his brother's instructions, but Robert appears to have lost his head and charged after the enemy so fiercely that he and his men reached the Egyptian camp before the alarm could be raised. Here was justification indeed! His brother's aim was to destroy the camp and Robert decided he could achieve this on his own. If he had now waited for the rest of the army to catch up he might even have been able to vindicate his reckless disregard for orders. However, Robert now made the serious error of assuming that the fleeing Egyptians were the best the Egyptian army could offer. Eager to win all the glory for himself he ordered his division to charge into the nearby town of El Mansura.

Not all of Robert's companions shared his impetuosity. The English knight William of Salisbury tried to persuade him to wait for his brother, but Robert taunted him with being a coward. This was an insult that no knight could accept and so William grimly joined the Crusaders as they rode into the narrow streets of El Mansura. Here the knights were at the mercy of the Muslims, who attacked them from the rooftops with bricks and tiles. Meanwhile, the main Egyptian army, so far uncommitted to the fighting, moved into the town and cut off the Crusaders' retreat. Few escaped and Robert was killed, along with William of Salisbury, the Master of the Templars, and over three hundred knights. Through the impetuosity of his brother King Louis had lost a third of his cavalry at a single stroke.

The Springs of Cresson

Crusader contempt for the Muslims was at the root of many of the military reverses they suffered in the Holy Land. For sheer crassness it is difficult to find a better example than that of the Master of the Templars, Gérard of Ridefort, at the Springs of Cresson in 1187. Charged with a diplomatic mission, Gérard was travelling towards Tiberias with a small group of companions. When he heard of the presence in the locality of a Saracen army of 7,000 mounted warriors, led by Saladin's general, Keukburi, he immediately collected as many Christian soldiers as he could find at short notice. With just 140 knights and 300 footsoldiers he decided to attack Keukburi, who was blissfully unaware of Gérard's presence. In spite of the strongest arguments of Roger de Moulins, the Master of the Hospitallers, that the disparity of numbers was too great, Gérard was not deterred. He asserted that the superiority of Christian knights was such that victory was certain. His own deputy, the Marshal of the Templars, James of Mailly, sensibly recommended retreat, but Gérard accused him of being a coward and of loving his blond head too much to risk losing it. James retaliated by saying that he would die in battle, a brave man, but that Gérard would flee like a traitor – a prediction that was to prove correct. Gérard was so impetuous that he charged the Saracens without even waiting for his infantry, who played no part in the fight. The knights were first surrounded and then massacred, with James of Mailly and Roger of Moulins among the dead, but Gérard escaped with two other Templars. As if to prove that a fool cannot mend his ways, Gérard played a similarly disastrous role in the much more significant and catastrophic battle of Hattin later the same year (see p. 56).

'We shall always win by reason of pluck: and, if it is not the only cause of victory, it is always the most essential factor and the one without which we cannot hope to succeed.'

Major Douglas Haig, 1896

This early quote indicates Haig's belief that courage alone can win battles. It contrasts with the more scientific thinking in France and Germany.

Nicopolis

The French knights at the battle of Nicopolis in 1396, during the last Crusade, combined impetuosity with a contempt for both their Hungarian allies and their Ottoman Turkish enemies. This already potent cocktail was topped off with a strong dose of arrogance which made them disobedient and completely reckless. All the elements necessary for a military disaster of the first magnitude were present.

Such was the cocksure arrogance of the French, they maintained that 'if the sky were to fall we would uphold it on the points of our lances'. Dismissive of their opponents and ignorant of their tactics, they fought according to their own prejudices and preconceptions. Whenever King Sigismund of Hungary counselled caution they took it as an insult to their courage. Even when he told them that he intended to send in a screen of light infantry ahead of them to clear the way for their charge, they declared that they had not travelled so far to be preceded into battle by a rabble of peasants. The French leader, Constable D'Eu, spoke for them all when he declared, 'to take up the rear is to dishonour and expose us to the contempt of all'.

On the day of the battle, Sigismund made one last attempt to reason with his allies, but they replied that he was simply trying to deprive them of the glory. With that D'Eu led out his knights to attack the Turks. As Sigismund had predicted, some of the strength of the French charge was absorbed by the light infantry the Turks employed in front of their main army. Nevertheless, the French cut through them, then proceeded to take out a line of archers. Feeling that things were going well, the French leaders now rested for a moment. However, hopelessly unaware of Turkish tactics, they did not know that they had so far only broken the Turkish advance guard. The main body of Turkish cavalry now swept through the tired French knights, killing some and capturing the others. Sigismund was later to reflect on his defeat, 'We lost the day by the pride and vanity of these French; if they had believed my advice, we had enough men to fight our enemies.'

The Timid

There are times when indecision can be as dangerous as making the wrong decision. The commander who just lets things happen because he is afraid of making a mistake is almost certainly already making one. There have been generals who have let their fear of responsibility dominate their thinking to such an extent that they have been defeated before the enemy has even fired a shot. Such a man was the Austrian general Karl Frey von Lieberich Mack.

'The unhappy General Mack'

Mack's speciality was planning campaigns rather than fighting them. Had he known the precise details of the strategy by which Napoleon surrounded him at Ulm, though defeated he might have felt honoured to be on the receiving end of such supremely skilful manoeuvring. Where Napoleon used speed, secrecy and weight of numbers, the Austrian lost himself in bureaucratic detail and fanciful imaginings, in which he convinced himself that, as if by magic, the English had invaded France via Boulogne and that Napoleon was fleeing into Germany pursued by a vengeful Britannia. The hard evidence for this extraordinary assumption was that an Austrian agent had overheard a conversation in a French village inn.

In a way Napoleon's masterly campaign was wasted on Mack, who proved disappointingly passive throughout. Even though he and the Archduke Ferdinand of Austria commanded 70,000 troops on the Danube, they did little more than blunder into the trap that Napoleon set for them. Mack's behaviour at Ulm has been likened to that of a rabbit hypnotized by a snake. Yet had he been able to link up with the advancing Russian troops of General Kutuzov, the combined Austro-Russian army would have posed a real threat to France. This Napoleon could not allow, and he set out to destroy Mack before the Russians arrived.

Mack's decision to concentrate his forces in Ulm played into Napoleon's hands and his failure to reconnoitre the advancing French columns meant that nothing was done to counteract French strategy. As belief in the English invasion faded, relations between Mack and the Archduke grew strained. Ferdinand insisted that there should be a strike against one of the approaching French columns, or at least a defence of the Danube crossings, to delay the French and give the Russians time to arrive. But Mack's timidity overcame him completely and little was done. Even Napoleon was surprised at this – he had assumed that Mack would try to fight his way out of the trap. Ferdinand now launched several feeble attempts to break out, even failing with 25,000 men to over-

The 'unhappy General Mack' – defeated by his own indecision and timidity, rather than the French Grand Army – surrenders to Napoleon at Ulm.

come General Dupont's single division of 4,000 at Haslach. Despondently, Mack called his troops back to Ulm, much to the fury of Ferdinand. Mack insisted on keeping his troops concentrated inside the city, but the Archduke knew that unless they broke out now they were lost. Despairing of convincing his fellow commander, Ferdinand made a sortie with 6,000 cavalry, leaving Mack to his fate. But even now the Austrians did not get far. They were pursued by Marshal Murat, who took prisoner not only the Archduke but two further Austrian divisions, totalling 26,000 men.

Inside Ulm itself Mack was in despair. Vainly clinging to the hope that the Russians might arrive in time to save him, he called on the French for an armistice. Napoleon, who knew that Kutuzov was still a hundred miles away, was happy to indulge him. But when news of Murat's success in capturing the Archduke Ferdinand reached him, Mack's resistance collapsed completely and he raised the white flag, five days before the time allotted. As he told Napoleon, 'Sire, here is the unhappy General Mack.' With him he surrendered 27,000 men, to add to Murat's impressive haul. At virtually no cost to himself, Napoleon had completely destroyed the strongest of the Austrian armies sent against him in 1805, with not one man in twenty in the *Grande Armée* firing a shot in anger. Mack's tragedy was that he feared to fail, and in the end failed without trying – an unforgivable crime for a military commander. He was court-martialled and sentenced to be shot by the Austrians, though the Emperor reprieved him and he spent ten years in prison instead.

Marooned on the French coast

Sometimes a job is too big for a man or, at least, he thinks it is. This was certainly the case with the 70-year-old British general Thomas Bligh in 1757, during the Seven Years War. He became the scapegoat for a series of bungled attacks on the French coast dreamed up by the prime minister, William Pitt, first Earl of Chatham. Rather than use British troops to help the Prussian King Frederick II in Germany, Chatham believed he could tie down French troops by staging hit-and-run attacks on French coastal towns. The attacks often ended in farce and sometimes in disaster. Humiliation followed an attempt by the Duke of Marlborough (the grandson of the great first duke) to evacuate troops from St. Malo when he saved his men but left his table silver behind. An obliging French commander returned the Duke's silver once he was sure the British had pulled out for good.

Bligh's first attack was on Cherbourg in August 1757. He captured the port, burnt French shipping and carried off brass cannons, which were triumphantly paraded through London. However, his second trip, to St. Malo, was a very different affair. After he had landed his troops on French soil a squall developed, foring the fleet to look for shelter at St. Cast, several miles down the coast. Bligh was now cut off from his transport and seemed at a loss to know what to do next. French troops were beginning to appear and several small clashes took place. Bligh ordered his men to march down the coast towards St. Cast so that he could re-embark. But first he needed to cross the River Équernon. That Bligh was a landlubber cannot be doubted, for he proposed a crossing at high tide. When this proved impossible he merely sat and waited some nine hours until it was possible for his men to wade across, their passage now hampered by heavy fire from the growing numbers of French troops. Instead of forming up his troops, some 10,000 strong, and driving off the French in order to win himself some time, Bligh did nothing and even allowed armed peasants to hem in his force. Declaring it was too late to re-embark that day, he made camp. Meanwhile, more French troops continued to appear, with a further force of 10,000 expected to arrive from Brest at any time.

In the morning Bligh continued his march to St. Cast, but by now discipline was breaking down among the British troops. Instead of inspiring them with confidence Bligh made it clear to everyone that he was afraid they might all be trapped. A message was sent to the fleet that the troops would be evacuated early the following morning. It might be supposed that Bligh would have tried to slip away quietly rather than raise the whole countryside with a vigorous reveille, but the British assembled on the beach to a noisy background of drumbeats. In spite of the obvious need for haste Bligh marched the men off in single column, thus delaying re-embarkation by several hours. But this time his luck had run out. The large French force from Brest arrived and now attacked the retreating British soldiers as they took to the boats. Eventually the Guards Brigade, under General Drury, was forced to counterattack the French to allow the embarkation to continue. But it was a vain sacrifice, for French artillery splintered many of the boats and some men were drowned while others lost their equipment. Finally, the Guards were overrun by the French and Drury killed, along with over 750 officers and men.

Though everyone liked Bligh, 'a very good-natured man', he was clearly not up to the rigours of amphibious landings. Even when, on the day of re-embarkation, he was confronted by an equal number of French troops he was too timid to fight them. Lord Ligonier, the British commander-in-chief, was horrified at Bligh's timidity and at the ruin of the Guards brigade. Bligh escaped a court-martial for his conduct of the operation, but his career was ruined. As a final insult, when he went to pay his respects to King George II, the monarch cut him dead. It was not just the end of the general's career, it was the end of Chatham's strategy of hit and run, which was years before its time and needed men of a calibre far beyond the amiable Thomas Bligh.

Pinkerton's Private Detective Agency

To have a reputation for excellence without ever having to prove it is good fortune indeed. Such was the enviable position of the 'Redoubtable McC', as the Confederates unkindly called Union General George McClellan during the American Civil War. 'McC' was the very model of a modern major general. He looked splendid, he had the best-dressed staff of any general on either side, and his manoeuvres – on paper – were Napoleonic in style and vision. Unfortunately McClellan was afraid to risk his reputation by actually putting them to the test. So cautious was he that President Lincoln referred to him as 'a stationary engine' and the whole of his Army of the Potomac as no more than 'George McClellan's bodyguard'. Lincoln was looking for a fighting general who could give him victories, but McClellan preferred to fight theoretical battles and so his victories were hypothetical ones. His over-estimation of Confederate strength became contagious and contributed to an inferiority complex among some Union commanders when facing legendary names such as Robert E. Lee, Stonewall Jackson or Jeb Stuart. Early in his command, McClellan had hoodwinked himself into thinking that the Confederates were holding Munson's Hill in force, only to discover that the guns which had held him in thrall were just logs painted black.

'My plans are perfect. May God have mercy on General Lee, for I will have none.'

General Joseph Hooker, before the battle of Chancellorsville, May 1863

The incompetent Hooker's plans fell apart. Lee and Jackson had made their own plans and inflicted a crushing defeat on 'Fighting Joe'.

McClellan had an unusual line in intelligence services. Dissatisfied with what the army could offer, he employed Pinkerton's Private Detective Agency to gather information about the enemy. However, as a private agency, Pinkerton's believed in keeping the customer satisfied by giving him what he wanted, rather than providing him with an objectively truthful account. McClellan wanted to hear that the Confederates outnumbered him heavily, because this would give him an excuse to delay doing anything. Pinkerton, either through incompetence or more probably through design, fed McClellan inaccurate and inflated figures. This allowed the 'Young Napoleon' to retreat even further into his shell.

In August 1861 McClellan informed Lincoln that the Rebels had four times his strength. In March 1862 Pinkerton reported Lee's 40,000 as 80,000. By June 1862, McClellan's 100,000 troops were being held up by McGruder's 23,000, yet he was able to tell the President that 'the rascals are very strong'. By October 1862, his estimation of

Lee's force at 150,000 would have given the Confederate commander the biggest army of the entire war. In sharp contrast was a report in *Harper's Weekly* by a French military observer that the Rebels had just 60,000 'ragged, dirty and half-starved' troops.

On one occasion the 'Young Napoleon' was actually presented with the entire Confederate battle plan, wrapped around some discarded cigars. This prompted him to attack Lee at Antietam Creek, but again timidity was his undoing. By holding back an entire corps, to cover against defeat, he robbed himself of the chance of a decisive victory. This was too much for Lincoln, who decided that McClellan had to go and began the search for a man who would fight. Unfortunately, Lincoln and the Union forces had to endure some even more dangerous blunderers than McClellan before the right man was found in Ulysses S.Grant.

The Bigoted

When based on generalizations about the national characteristics of his enemy, a commander's decisions are unlikely to be good ones. Stereotyped views of other nationalities are the product of personal or national prejudices and are unlikely to have any objective validity.

Singapore Slings

The idea that the Japanese were small, weak, had bad eyes and could only design inferior copies of Western products caused the British to underestimate them as adversaries. British contempt for the Japanese as opponents before the fall of Singapore in 1942 led to complacency in defending the island and eventually to disaster. In December 1940, Air Chief Marshal Brooke-Popham commented on some Japanese prisoners he had seen in China, 'I had a good close-up, across the barbed wire, of various sub-human specimens dressed in dirty grey uniforms, which I was informed were Japanese soldiers. If these represent the average of the Japanese army, the problems of their food and accommodation would be simple, but I cannot believe they would form an intelligent fighting force.' The same man, on receiving the suggestion that Singapore's air defences needed to be improved in the light of Japan's modern air force, commented, 'We can get on all right with the Buffaloes here. [The Brewster Buffalo was an antiquated American fighter plane hopelessly slower than the latest Japanese aircraft.] They are quite good enough for Malaya.' In fact, the British air force was simply overwhelmed by the modern planes and tactics of the Japanese.

On the ground things were little better. One British battalion commander inspecting his troops in Singapore asked, 'Don't you think they are worthy of some better enemy than the Japanese?' Another officer of a distinguished British regiment commented, 'I do hope . . . we are not getting too strong in Malaya, because if so the Japanese may never attempt a landing.'

When the Japanese Zero-sen naval fighter cleared the skies of the antiquated British aircraft considered good enough for a war against Japan, the cultural shock was enormous. It was widely assumed that they must have been built and piloted by Germans. But the notion of white supremacy received its final blow when it was revealed that the capture of Singapore's garrison of some 120,000 British troops had been achieved by a Japanese force of only a third that many men.

'I believe that defences of the sort you want to throw up are bad for the morale of troops and civilians.'

Lieutenant-General Arthur Percival, Singapore, December 1941

Singapore had been designed as a fortress, equipped with guns facing out to sea. It was not expected that an attack could come down the Malayan peninsula. When it was pointed out to the commander that he had no landward defences, he refused to build any, saying it might cause the people to lose confidence in the authorities. The result was that the Japanese encountered no effective line of defences at any stage in their conquest of Malaya and Singapore.

The Herero Wars

Underestimation of the enemy based on racial prejudice played a part in a number of military disasters in the colonial context. Even the normally competent Germans were not immune to the pitfalls of ethnocentrism. In the Herero Wars in Southwest Africa in 1904–5, the Germans suffered a series of setbacks because they failed to take their Herero adversaries seriously and tried to fight a war in a manner inappropriate to the surroundings.

Many German soldiers saw the war as an undemanding way of winning some glory and perhaps a medal. The reinforcements sent from Germany, called 'boys' by the Hereros, were quite unprepared for the difficulties of what was essentially a desert campaign. The officers, many of whom had had to use their connections in high places to get this chance, did not take things seriously enough, arriving with crates of vintage champagne, their favourite cigars and hunting rifles. No doubt they hoped that the war would not interfere too much with what was essentially an exotic holiday. They were in for a terrible shock. In a desert country with poor roads their modern weapons imposed problems of transport that had not been anticipated in Berlin. A machine gun uses ammunition weighing hundreds of pounds; artillery shells weigh even more: how could these essentials be transported around the country against a fast-moving enemy? Moreover, telegraph lines could not be protected and so the Germans had to fall back on heliographs. Things were not going to be as simple as they initially thought.

The fate of Major Glasenapp's column illustrates the problems the Germans faced against a skilful and elusive enemy. His force was entirely made up of new arrivals to the colony and even though this should have persuaded him to act cautiously, Glasenapp was not a man to hold back when glory beckoned. He decided to hunt down the Tjetjo Herero, who, he was informed, were in the vicinity of the town of Gobabis. Anxious to win renown Glasenapp ordered his 'rookies' to chase the tribesmen in exhausting conditions, in fierce heat and with water in short supply. Even averaging twenty miles a day, the Germans could not bring the Herero to battle. When eventually forced to stop at

Onjatu to allow his men to rest, he heard from scouts that the Herero were just a few hours ahead. On setting off again, Glasenapp encountered a large herd of cattle with just a few Herero herdsmen, who immediately ran away. The Germans seized the herd, but as they rode on the enormous dust cloud produced by the cattle prevented them from seeing where they were going. They had been passing through an area of thick bush and, oblivious to the danger of an ambush, came out into a clearing in some disorder. The Herero were waiting for them and immediately opened fire from the bush on the other side of the clearing. The Germans formed up into a ragged line and fired volleys into the bush with about as much success as Braddock's men on the Monongahela in 1755 (see p. 86). The German machine gun raked and peppered the bush to little effect, and the Herero simply shot down every man who tried to operate it. Eventually Glasenapp had to call off the battle and retreat to the bush, leaving behind 26 German dead and carrying many wounded with him. The Hereros suffered no casualties and simply slipped away into the bush to continue their guerrilla war.

It is perhaps as well to add that the Germans eventually suppressed the Herero tribesmen by a policy so severe that it virtually amounted to genocide.

The Unsure

There have been many soldiers who, when offered command, have not felt competent to do the job. In such cases this has not prevented them from eventually taking up their posts, willingly or otherwise, and attempting to do their best. This lack of self-confidence seemed to be most apparent in the Habsburg armies of Emperor Franz Josef, which, in the 1859 war with France and the decisive war against Prussia in 1866, were plagued by commanders who were afraid of responsibility and allowed fear to dominate their thinking. The result was two disastrous – yet avoidable – defeats.

When in doubt do nothing

In 1859 Emperor Napoleon III of France, in alliance with the Italian kingdom of Piedmont, was fighting the Austrians for control of Lombardy. The sides were evenly matched, with certain advantages, notably logistical ones, resting with the Austrians. Unfortunately the Austrian commander, Count Gyulai, had never held field command before and asked unsuccessfully to be relieved on the eve of the campaign.

When hostilities began between the Franco-Piedmontese and the Austrians, Gyulai had a great chance of preventing the two enemy armies from linking up and could have beaten them one at a time. Outnumbering the Italians by two to one, he should have immediately marched on their capital of Turin, traversing the flatlands between the city and the Ticino River. The Emperor, Franz Josef, ordered him to act, but Gyulai did nothing.

Against Gyulai's five army corps and a cavalry division, the Italians could only muster five divisions and a cavalry brigade. Everything pointed to an overwhelming Austrian victory. Morale among the Austrian soldiers was high, their army having won confidence-boosting victories against the Italians, including two as recently as 1848–9 at Custozza and Novara. All that was needed was for the commander to do something positive. When Gyulai did act it was simply to march up and down on the Austrian side of the Ticino River, before slowly crossing it and then ineffectually staying put again. But now the weather changed and unseasonable rain turned the whole area into a quagmire. This gave Gyulai an excuse for his slow approach towards the city of Novara. Suddenly, to the amazement of everyone, he stopped the advance and ordered a retreat back to the positions from which they had started, across the Ticino.

When the Austrian army came off worse in an encounter with the French army at Magenta, Gyulai panicked at the setback, evacuating most of Lombardy, including Milan, and retreating a hundred miles. This inflated the defeat at Magenta out of all proportion, inspired nationalist feelings throughout Italy and destroyed Austrian morale. The Emperor, Franz Josef, now took over command of the army, and Gyulai accepted his dismissal with pleasure. The officer who told him the Emperor's decision reported that he was 'in the best of spirits. He sees nothing wrong at all. He has all his comforts, good cooking, cards after dinner. He invited me to dine, but I excused myself . . . this HQ turns my stomach and I could weep.'

After the defeats by the French in Italy in 1859 and the loss of Lombardy, there was a 'deep spiritual malaise' in the Austrian command, as if they expected to lose every battle before it was fought. The desire to avoid responsibility was coupled with a sluggishness and a virtual death wish. In 1860 General Degenfeld-Schönburg tried to refuse his appointment as War Minister, while in 1864 Henikstein declared that he did not have the right qualities for a chief of staff. He declared, 'I am not the man for this important position. I lack the ability, the knowledge and therefore the necessary self-confidence.' He was appointed against his wishes and proved his incompetence during the campaign of 1866 (see p. 60).

The Austrian Bayard

The one Habsburg general to emerge from the war of 1859 with his reputation enhanced was the Magyar Ludwig Ritter von Benedek. He was loved by his men and widely respected in his profession, being called 'the army's Bayard' – after the fearless and irreproachable sixteenth-century French soldier – and 'the restless and indefatigable Benedek'. Yet he was more aware of his limitations than others. As a corps commander he was excellent, but the control of an army of 150,000–200,000 men was beyond him. He needed to be able to see what was happening and react to it rather than plan a battle in advance with maps. His choice as commander-in-chief in 1866 seemed a natural one, but in fact it was to prove fatal to the Austrian cause in the war against Prussia. He was reluctant to accept the promotion and seemed to lose the energy and drive for which he had been renowned. Although only 62, he was suffering from recurring bouts of illness, and feared his health might not hold up in prolonged campaigning. He also disliked moving his command from Italy, where he had done so well, to central Europe. 'So I am now supposed to study the geography of Prussia! What do I know about a Schwartzer Elster or a Spree? How can I take in things like that at my age?'

Nor had Benedek any reason to be confident in the capacity of some of his specialist staff. When one Lieutenant-Colonel Beck was preparing a geographical study of central Germany he asked one of his superiors for the maps which that officer was supposed to have collected while in Germany. His senior, who had preferred to spend his time in casinos instead of carrying out surveys, told him to find what he wanted in a Baedeker.

Benedek seemed bowed down by responsibility. In choosing two chiefs of staff, where one was normal, he revealed confusion and indecision in his thinking. The two men chosen, Krismanic and Henikstein, were an ill-assorted pair who worked badly in harness. Neither was popular nor known for his efficiency. As a fellow officer wrote, 'My over-learned but lazy friend Krismanic and my unlearned friend Henikstein, who is as fit for his place as I am for composing an opera, appear to have bungled.' Krismanic was conservative and unduly timid, and so cynical was Henikstein in 1866 that even before the campaign began he recommended to Benedek that Austria's main effort should be to fortify Vienna for when the retreat became necessary.

The timidity and pessimism of these two gloom-laden advisers disheartened the normally daring and confident Benedek. Defeat seemed to be in the very air that he breathed. At the battle of Könnigrätz, Benedek adopted an entirely passive strategy towards the two main parts of the Prussian army that were moving against his defensive position on the heights north-west of the town. The Prussian commander, Moltke, was taking an enormous risk in separating his forces in the way that he had, but his confidence in the ability of his commanders to coordinate their attacks was well founded. Nevertheless, had Benedek attacked either of the approaching Prussian armies with his full strength, the Austrians would surely have won a crushing victory, with odds of two to one in their favour offsetting the advantage the Prussians had in their needle gun. Ironically, it was only in the brilliant disengagement of the beaten Austrian army after Könnigrätz that Benedek seemed to have regained his confidence.

'The Turk is an enemy who has never shown himself as good a fighter as the white man.'

Staff Officer before the April landings in Gallipoli, 1915

This prejudiced and misleading observation encouraged the Allied soldiers to underestimate the fighting qualities of their opponents. In fact, the Turks at Gallipoli under good German leadership, or led by Mustapha Kemal, fought well and won the respect of the British and Anzac troops.

THE BATTLE OF PLISKA (811)

Khan Krum of Bulgaria, with his frequent raids on Byzantine territory, was a constant thorn in the side of the Emperor Nicephorus I. In 811 Nicephorus decided to put an end to the Bulgar threat once and for all. He raised an army of 70,000 men, including the cream of Byzantine chivalry. It was an army glittering with the gaudy accoutrements of most of Constantinople's high-ranking courtiers, as Nicephorus thought it wise to take

with him on the campaign anyone who might plot against him in his absence. With him went his son Stauricius, as well as the whole Tagmata – the Byzantine standing army – and units from Asia Minor and Thrace. It was an army strong in every department, notably in its ability to besiege fortified cities. To the populace of Constantinople, who watched it depart, it seemed that nothing could resist its awesome power.

The Bulgar Khan was thoroughly overawed when he heard of the force the Emperor was sending against him and immediately sued for peace. But Nicephorus, having gone to the trouble of raising such an army, was unwilling to disband it while Krum still posed any threat. Even as the

Bulgars panicked and Krum withdrew his troops to the north, Nicephorus marched irresistibly towards the Khan's capital at Pliska. Sweeping down on the undefended wooden town, the Byzantines burned Krum's palace and killed over twelve thousand of the town's inhabitants. Krum had ordered a relief force to try to save Pliska, but Nicephrous destroyed this force so completely that it seemed as if it had never existed. The Byzantine troops, replete with loot from Krum's treasury, had rarely had such an easy and successful campaign in Bulgaria. When news of their success reached Constantinople the people rejoiced in the belief that Nicephorus had divine approval for his actions.

Yet although Nicephorus had destroyed Krum's capital, he had not destroyed his army; and the destruction of the Bulgarian army was essential if the campaign was to have any permanent value. In the mountains north of Pliska, Krum was raising fresh troops, hiring mercenary Avars, and even arming women. The Byzantines still had much to do.

Knowing that he could never face the Byzantine army in open battle, Krum decided to try to use his familiarity with the terrain to trap Nicephorus, whose success so far might, he thought, make him careless and overconfident. In this he was right. Convinced that his march through Bulgaria was little more than a procession, Nicephorus began to neglect normal precautions, particularly the need to scout ahead. As a result his army walked into one of Krum's traps in a steep-sided valley north of Pliska. Nicephorus's massive force marched on unawares until riders from the vanguard brought Nicephorus the startling news that the exit from the valley had been blocked by the Bulgars with a palisade of logs. Even before the decision to retreat could be given, riders from the rear reported that Bulgar troops had appeared in the entrance to the valley and had held off Byzantine attempts to prevent them building another log palisade of immense strength. The Byzantines were trapped.

Everyone now awaited the Emperor's decision. The Byzantines vastly outnumbered the Bulgars, but in the narrow spaces of the valley Nicephorus would not be able to employ his mailed cavalry to advantage. The cliffs on each side of the valley made escape that way quite impossible. The only alternative was to force a way through one of the palisades, whatever the cost. Faced by this dilemma Nicephorus lost confidence completely and could think of nothing to do but set up camp. His generals pleaded with him to lead an immediate assault before his troops realized their predicament and panicked, but Nicephorus refused. Even when his officers persuaded his son Stauricius to reason with him, the Emperor refused to do anything, saying, 'Even if we become winged, no one could hope to escape destruction.'

For two days Nicephorus and his army camped in the valley, making no attempt to break down the palisades, even though they were equipped with siege weapons. The Emperor had become fatalistic and almost abandoned control of events, preferring to wait and see what would happen. This was exactly what Krum was hoping for, and on the third night he ordered his men to bang their shields and shout news of their hopeless position down to the Byzantine troops. Having thus demoralized his enemy, Krum now decided to attack. Aiming first at the camp of the Tagmata and the royal tents, the Bulgars suddenly swept down from the hills, taking the Byzantines by surprise. The Byzantines were quickly overwhelmed, and Nicephorus himself was killed in the fighting, the first Roman Emperor to die in battle since Valens at the battle of Adrianople in 378.

In their panic many Byzantine soldiers were driven into a marshy riverbed, where they were drowned or massacred by the Bulgars. Eventually, the river bed was so replete with corpses that thousands of men crossed it on the bodies of their colleagues. But those who passed the marshes still had the log palisade to contend with. Some Byzantines climbed the barrier only to fall to their deaths in the deep ditch on the other side. Others set fire to the logs and rode through the flames, only to fall into the same concealed ditch. Once the logs had thoroughly burned, the pitiful remnants of Nicephorus's great army managed to make their escape, still pursued by the triumphant Bulgars.

In the Emperor's camp the élite of the Byzantine military system had been wiped out. With his father dead, Stauricius had little time to reflect on his elevation to Emperor. Severely wounded, he and his guards managed to cut their way clear and escape, though for Stauricius death was to come six months later from the lingering effects of his wounds. Nicephorus's head was hacked off and displayed on a pole for a few days to the delight of Krum's army. But the Bulgar leader had a more significant use for this symbol of his victory. The skull was eventually mounted in silver and used by the Khan as his drinking cup.

When news of the disaster reached Constantinople the people were stunned. How was it possible for such a disaster to occur after so successful a campaign? To many people it was clear that God had withdrawn his favour from Nicephorus. To others it was hubris: Nicephorus had grown too great and too successful – and too careless.

In any case, the Byzantines could not tolerate the humiliation when they heard of Krum's ghastly taste in drinking cups. The Empire must have its revenge. The following year, the Emperor Michael, who had succeeded the short-lived Stauricius, heard from two Christians who had escaped from Bulgaria that Krum was planning to invade Byzantine territory. Leading out a great army, Michael marched to Mersinicia, where Krum was encamped. For fifteen days the two armies remained within a few miles of each other, neither daring to begin the assault even though the Byzantines are reported to have outnumbered the Bulgars by ten to one. Michael resisted every effort by his officers to persuade him to attack Krum, fearing that he might – like Nicephorus – be drawn into a trap. Eventually, the commander of the left wing, John Aplaces, managed to persuade Michael to strike while there was still a chance. As John led his troops forward the Bulgars immediately turned tail and fled. Encouraged, John charged onwards, expecting to be supported by the rest of Michael's army. However, they too had fled, leaving the left wing completely alone. Krum's army now turned round, encircled John's troops and annihilated them. Seeing the disaster, Michael fled back to Constantinople. He was the victim of treachery from within his own army. Had he acted sooner he would have achieved a total victory, but by delaying for over a fortnight he gave Krum time to come to an agreement with dissident officers in the Byzantine army. On reaching the capital, Michael was forced to surrender the crown to one of his generals, Leo the Armenian. It was Leo who had led the dissident officers, and it was he who had engineered the flight of the Byzantine army.

THE BATTLE OF TRENTON (1776)

Johann Gottlieb Rall, Colonel of one of the Hessian regiments the British used as mercenaries during the American War of Independence, was a reckless, hard-drinking bully. As a professional soldier he should have known better than to allow his personal prejudices to affect his judgement, but he hated rebels. He did not understand their cause and he did not respect them as fighting men. As commander of the city of Trenton, New Jersey, in December 1776, the Hessian's fatal underestimation of the American soldier was not only to cost him his life but also revitalized the American Revolution at a time when few – perhaps not even George Washington – felt it could survive much longer after a string of heavy defeats.

Rall was not a man to take advice. He weighed up a situation according to his own limited percep-

George Washington, the cautious but inspiring commander of American forces and the first president of the USA. At the end of 1776 Washington's forces were in dire straits and the collapse of the revolt of the colonies seemed only a matter of time. But unexpected victory at Trenton breathed new life into the rebel cause.

tions and stuck obstinately to his view. When his junior officers advised him to fortify Trenton against attack he sneered that he needed no earthworks against such men as the Americans, and that the bayonet would do. Rall's confidence meant that he listened to no one, never visited the pickets to see that they were correctly positioned and cared little for the welfare of his men. As long as his men drilled and marched and the band played then Johann Rall was happy as a lark. Meanwhile, he intended to enjoy himself. He caroused with his friends till the small hours, slept late in the morning, and kept the men on parade in the snow while he took his bath in comfort.

Yet Rall's confidence was only skin deep. Below the thin veneer of military 'spit and polish', he was a worried man. News of rebel raids all around Trenton made him edgy. He felt exposed and liable to attack. Yet he could not admit to weakness in front of his men and to have fortified the city might have suggested that Johann Rall could be rattled – an unthinkable notion. On the bitterly cold, snowy evening of Christmas Day, Rall received a letter from a American loyalist warning him that an attack was imminent. But Rall was too busy drinking and playing cards, and put the letter unread into his pocket. This was a fatal mistake.

By December 1776 the revolt in the American colonies was fading. Defeats at Fort Washington and Fort Lee had caused Washington to abandon New York and retreat into Pennsylvania to guard Philadelphia, the rebel capital. Assuming that there would be no further fighting during the harsh American winter, the British commander, Sir William Howe, sent his troops into winter quarters. But Washington knew that he needed a victory to restore confidence in the rebel cause. He therefore decided to cross the partly frozen River Delaware and attack the city of Trenton, held by the 1,400 Hessian soldiers of Colonel Rall. His force of 2,400 men crossed the river and at dawn on 26 December 1776 took the Hessian garrison of Trenton by surprise.

The reason the Americans had achieved complete surprise was that Rall's second-in-command, Major von Dechow, had decided to cancel the regular dawn patrol because of the bitter cold. Those Hessians who were on picket duty had their backs turned to the biting north wind and were surprised by unseen enemies. The first any of the Hessians knew of an attack was when an officer, hearing

shots, went to check on the dawn patrol, only to find that they were comfortably warming their toes indoors. He then hurried to beat on Rall's door, but the Colonel, still in bed, was not to be rushed and sent him packing. Before the hastily assembling Hessians, groggy from their Christmas celebrations, could respond they were raked with grapeshot from American cannons. Poor Rall, emerging in haste from his warm bed, seemed bewildered to find himself surrounded by Americans. Hastily summoning the regimental colours and, most important of all for Rall, the regimental band, he organized his men in counter-attacks. To the sound of bugle and fife and drum, the Hessians, some of them half-dressed, moved in regular lines towards the Americans, who shot them down in droves. Soon the well-ordered ranks gave way and the men fled towards a bridge, which was still in Hessian hands. Rall, now mounted and shouting and waving his sword about his head, was hit by a shot and immediately became less voluble from loss of blood. He just had time to order a retreat before slumping into unconsciousness but did not live to see his regimental colours, along with nearly a thousand of his men, surrendered to the Americans, whom he had been so foolish as to underestimate. American casualties were reported at just five.

General Howe was horrified. He could not believe that 'three old established regiments of a people who make war a profession should lay down their arms to a ragged and undisciplined militia'. From this it is clear that Rall was not the only officer on the British side who needed to learn respect for his American foes. It would take another six years of fighting before the lesson was truly learned. Washington's success at Trenton, and Rall's failure, was to give the Americans the chance to administer that lesson.

THE BATTLE OF MAJUBA HILL
(1881)

General Sir George Pomeroy Colley was one of the most brilliant officers to pass through the army Staff College in the nineteenth century. In fact, he completed the two-year course in less than half the time allowed and scored the highest marks ever recorded. He was well-read, a gifted linguist, a talented painter, and an engineer and scientist of note. He had seen action in the China War of 1860, served in India as a colonel and in South Africa as chief of staff to Lieutenant-General Sir Garnet Wolseley. In 1880 his skills as an administrator won him appointment as Governor of Natal and High Commissioner for Southeast Africa. At 46, he was the brightest star in the British army. However, what happened to him in the last two months of his life served to undo a lifetime's achievements.

It has to be said that Colley was not well served by his subordinates. In the First Boer War Colonel Lanyon, Administrator of the Transvaal since 1879, gave Colley a completely inaccurate assessment of Boer military strength:

> I don't think we shall have to do much more than show that we are ready, and sit quiet and allow matters to settle themselves . . . They are incapable of any united military action, and they are mortal cowards, so anything they may do will be but a spark in the pan.

Lanyon, and through him, Colley, were about to make the cardinal error of underestimating their enemy. The British had just 1,760 troops to enforce their rule in Transvaal and once the Boers declared their independence they were bound to be badly outnumbered. On 20 December 1880, a 'battle' took place at Bronker's Spruit. A column of 264 soldiers from the 94th Regiment was stopped by a Boer commando, 1,000 strong, entrenched in high ground on all sides of the British. The Boers politely ordered the British to go back and gave their commander, Lieutenant-Colonel Anstruther, two minutes to make up his mind. Even though in a totally hopeless position Anstruther decided to fight it out and the column was decimated, 77 men being killed and over a hundred wounded. The Boer sharpshooting was astonishing; one soldier was hit eighteen times and every British officer was hit at least once.

The Boers now assembled 2,000 men at Laing's Nek, controlling the route from Natal into the Transvaal, and defied the British to do anything about it. In January 1881, with 1,000 British troops, Colley decided to invade the Transvaal. In view of the fact that he was outnumbered two to one and fighting against an enemy strongly entrenched in terrain he knew well, it was an

The British defeat at Majuba Hill in 1881 was as unexpected as it was decisive. Commanded by Sir George Pomeroy Colley, Britain's most able general, the British regulars proved no match for Boer farmers and schoolboys.

extraordinary decision. Colley must have been depending on the accuracy of Lanyon's assessment of Boer ineptitude. If so, he was in for a shock. At the battle of Laing's Nek his infantry were asked to charge up a steep slope and then seize Boer positions at the point of the bayonet. In fact, by the time they got to the top they were exhausted, many sinking to the ground for a breather, only to be hit from two sides simultaneously by pinpoint rifle fire. British casualties were 160 out of 480 men involved, the Boers losing just 41 men. Colley now admitted that he had underestimated the Boers. But would he learn from his mistakes?

Colley was determined not to let his defeat remain unavenged. Furthermore, he was driven by an extraordinarily ambitious wife, who goaded him to act decisively before any peace negotiations could take place. He needed a victory to restore his reputation. As one officer wrote at the time: 'Another defeat will kill Colley.'

Although the Boer position at Laing's Nek was a strong one, Colley could see that the mountain known as Majuba Hill, 6,000 feet high, overlooked the Boer camp. Majuba was so high that the Boers had not thought to garrison its peak and Colley decided that if he could get men to the top he would force the Boers to evacuate Laing's Nek. On 26 February, he led a force of over five hundred British soldiers on a night march to the top of Majuba Hill. Each man was heavily laden with extra ammunition and three days' supplies of food and water. After a difficult climb Colley found to his immense satisfaction that the hilltop was indeed deserted. Perhaps in his elation he let his guard drop, for from this moment everything went wrong, and much of it was the fault of Colley himself.

From the hilltop the Boer camp was just a mile and a half away, and the effect of overlooking the enemy in this way seemed to make both the commander and his men overconfident. Surprise was too vital an advantage to surrender lightly and Colley was greatly to blame in not preventing groups of Highlanders from waving and jeering at the Boers like holidaymakers or skylarking schoolboys. The Boers immediately opened fire on the tiny figures and achieved a hit, a prodigious shot at a range of 2,000 yards. From Colley's elevated position it seemed that the antlike activity in the Boer camp was just a prelude to an evacuation. Perhaps this assumption prevented Colley from giving the elementary instructions to dig entrenchments on the summit. Perhaps he did not believe the Boers could do what he had done and climb the hill. When junior officers asked if he wanted entrenchments dug, Colley simply replied that there was no need. Then he decided to have a nap.

Such complacency would have quickly evaporated if they could have heard what was going on in the Boer camp below. Hundreds of men were volunteering to climb the hill and chase the British off. Many were rejected; only 180 expert marksmen – mostly young farmboys – were allowed to make the climb. While 1,000 Boer riflemen set up a barrage of fire to keep the British heads down, the Boer assault party began its climb. It was this covering fire that a British officer reported as 'the Boers wasting ammunition'.

The young Ian Hamilton (later to command the disastrous British campaign in Gallipoli in 1915; see p. 159) ran to tell Colley that 100 Boers had reached the summit, but the general remained undisturbed. Even news that 200, then 300 Boers had appeared did not seem to bother Colley.

When Hamilton rushed back for the fourth time with an even more exaggerated report, he found that Colley had fallen asleep again. In fact, the hysterical Hamilton had given Colley the impression that hand-to-hand fighting was taking place, which was quite untrue. The Boers, some no more than 14 years old, had no intention of fighting three times their number of grown men. Instead, they stayed behind cover, and shot the Highlanders down as if they were shooting for the cooking pot.

Suddenly realizing the danger, Colley called up his reserves, many of whom like him had been sleeping and were half-dressed. They formed a firing line and fired a ragged volley which went wide. The Boers, who had dropped to the ground before the British fired, stood up and shot down 20 redcoats. Now many of the British soldiers were panicking and some began to scramble down the hill. While Colley coolly tried to rally his men he himself was shot through the head from close range and killed, it is claimed, by a twelve-year-old boy. What followed was unseemly chaos, with men surrendering or fleeing in panic.

The 554 British soldiers, under the army's most brilliant commander, had been routed by a group of boys and irregulars, whose military qualities had been disregarded and who had been called 'mortal cowards' at the start of the campaign. British casualties amounted to 93 killed, 133 wounded, 58 taken prisoner. Boer losses were one killed and five wounded. It was a thoroughly humiliating defeat. Many shared Queen Victoria's sentiments when she observed: 'Poor Sir George . . .'

THE AMRITSAR MASSACRE (1919)

The movement for Indian independence gathered pace after World War I and was accompanied by acts of violence against European civilians reminiscent of the early days of the Indian Mutiny of 1857. One of the centres of civil unrest was the city of Amritsar in the Punjab, and in April 1919 Brigadier-General Reginald Dyer was sent there to restore order. What followed constituted a military blunder committed in peacetime.

General Dyer has sometimes been portrayed as a racially prejudiced killer of innocent people, but this view is misleading. Dyer was no more and no less prejudiced than many of the British authorities in India at the time, who were haunted by memories of the Mutiny. They supported Dyer every inch of the way, indeed some thought him too lenient. Dyer was simply the man at the 'sharp end' of the operation who did the killing, while others sat in government buildings or well-stocked clubs and issued paper orders. The British establishment was frightened by the growing street violence in the Punjab and felt they needed a tough man of action. Dyer was certainly that, although it should be noted that the early signs of the arteriosclerosis which was to kill him were already giving him a lot of pain. Add this to his already quick temper, stir in the fact that it was the hottest season for many years and one has the ingredients for a tragedy.

On arriving in Amritsar with just over a thousand British, Indian and Gurkha troops, Dyer was met by the Deputy Commissioner, who gave him a full account of the attacks on European civilians, notably the savage beating of a missionary lady, Miss Sherwood, by a murderous mob. In this age of Imperial pride one can imagine the bold Englishman's breast swelling at the thought of the indignity suffered by this white woman. The Commissioner had issued a proclamation that assemblies of more than four Indians would be fired on, and Dyer received orders from the Viceroy himself to the effect that if troops were used an example should be set. Dyer saw clearly the implications of the Viceroy's message. He needed to set an example in Amritsar which would demonstrate Britain's determination to stamp out civil unrest throughout India.

Feelings were running high among the Indians as well as the British. Dyer had the Commissioner's proclamation read throughout the city, but the response was less co-operative than he had hoped. The crowds were not cowed but reacted by jeering and shouting that the Raj was dead and that the British should get out of India. A meeting was called in the Jallianwala Bagh, an open space of about an acre surrounded by houses and walls. This was an act of open defiance and the Indians knew it. They had been warned that the British would fire on such crowds, yet that afternoon at least five thousand people assembled to hear speeches.

The British response was furious, even hysterical. One British surgeon suggested bombing the crowd from the air. The Lieutenant-Governor of the Punjab declared Amritsar under martial law and Dyer set off to deal with the crowds in the Jallianwala Bagh. Taking a fairly small force of infantry, 25 Sikhs and 65 Gurkhas, only half of whom were armed with rifles, he drove to the meeting accompanied by his two armoured cars. The entrance to the Bagh was very narrow and so the cars were left behind and the troops marched up an alley and then straight into the arena. Without hesitating Dyer ordered his men to open fire into the crowd. At first there was a shocked silence and then someone shouted, 'They are only blanks.' But they were not blanks, and as the blood flowed people screamed and panicked, looking for a way out. Every exit was blocked and the only way out was the way Dyer's men had come in. The soldiers fired indiscriminately and once their magazines were empty they reloaded and kept firing. For some ten minutes they kept up the killing, firing altogether some 1,650 rounds. It is impossible to believe that any of these were wasted, so thick was the heaving mass into which they were shooting. Although Dyer claimed he had not seen any women or children, there were certainly many among his victims. Despite total casualties of 379 killed and over a thousand wounded, the British made no effort to help anyone, simply marching off the way they had come. Dyer had no doubts about what he had done and his decision was immediately endorsed by the Lieutenant-Governor. Among the British community Dyer was viewed as a hero and praised for his moderation.

Until this point Dyer could claim he was simply following orders – the universal cry of the concentration camp commandants and of military butchers throughout history. But now he lost all restraint. Always contemptuous of the Indians he now instructed that when meeting Europeans they should salaam. Anyone failing to do so was to be flogged on the spot. Indian bicycles were confiscated, along with other forms of wheeled transport. In a perverse attempt to strike at Indian intellectuals such as Gandhi and Nehru, Dyer ordered 93 Indian lawyers to work as coolies and forced them to watch public floggings.

It is doubtful if Dyer's mind was quite balanced during this period, though he was upheld by the vicious prejudices of the other Europeans in

Amritsar. Because of the attack on the missionary, Miss Sherwood, Dyer ordered anyone wishing to use the lane where the attack took place to crawl down it on hands and knees. Anyone refusing to do so was promptly flogged. The spate of flogging spread to other parts of the Punjab, culminating in an extraordinary incident in which all the members of a wedding party at Kasur were flogged for breaking the curfew.

When news of all this reached England the reaction was one of horror. Winston Churchill denounced Dyer in the House of Commons for this 'monstrous event'. The Hunter Committee was immediately set up to investigate the Amritsar Massacre, as it was now known by a shocked world. No other event could have served so to crystallize Indian public opinion. After six weeks the Hunter Committee issued its report which condemned Dyer for firing on the crowd without warning and continuing to fire until he had exhausted his ammunition. Dyer was relieved of his command and ordered back to England. To the eternal shame of the British people, Dyer was received as a hero, and a public subscription raised over £25,000, making him a rich man.

If Dyer's action was designed to restore and maintain British control over India, its effect was short term. In the long run it was a terrible mistake, convincing Indians that there could be no real equality between the races in India while the British remained and that nothing short of political independence would do. Only the supreme example of Gandhi kept the move to independence even relatively peaceful in the period before World War II.

THE BATTLE OF DIEN BIEN PHU
(1954)

The French defeat at Dien Bien Phu in 1954 was a result of military arrogance on the part of French commanders. They believed that their superiority over the Nationalist Vietminh – who were fighting for the independence of Vietnam from France – was so complete that they could defy common logic and defeat their enemy after conceding almost every advantage to them. An American adviser told the French, 'The Vietminh have no vehicles and no airplanes. How can they be mobile?' Such contempt for an Asian enemy was typical of the ethnocentrism of commanders schooled in the colleges of Western Europe and America, rather than the hard school of guerrilla warfare in the jungles of Indo-China.

General Salan had called the Vietminh commander General Vo Nguyen Giap, 'a non-commissioned officer learning to handle regiments'. It is therefore hardly surprising that when General Henri Navarre was appointed to succeed Salan in command of French forces in Vietnam in 1953, he viewed his opponent with equal contempt, as a peasant and an amateur. Navarre could not believe that Giap could assemble enough troops to defeat the French, nor supply those that he did bring together with the necessary food and equipment. If Giap faced the French in a 'stand-up' fight he would be crushed. In any case, the superiority of French artillery and air power was unquestionable. French views of Vietminh artillery are illustrated by Navarre's comment, 'They must have a gun or two, but most of the time the shells don't even explode. It's a farce.' Yet in each of these assumptions Navarre was to be proved wrong, with disastrous results for himself and for France.

From the start Navarre's decision to fight a decisive battle at Dien Bien Phu was based on a misunderstanding of his government's policy. By 1953 the strategic position in the Far East had changed. With Stalin now dead, the Soviets were trying to establish better relations with the West and saw a settlement of the Korean and Indo-Chinese wars as a way of achieving this. Now that the war in Korea was over, Communist China was in a position to send supplies to the Vietminh on a much larger scale. Both these factors were pushing France in the direction of a settlement of the Indo-China situation, which made Navarre's plan for complete victory quite inappropriate. France was heartily sick of the fighting and French politicians had no stomach for extending a war which might now involve Chinese intervention. Navarre therefore had been appointed to stabilize the situation prior to a negotiated settlement, not to risk everything on one 'stand-up' fight.

But Navarre began to 'paint a picture' of the situation which suited his own preconceptions, but was not based on fact. After he had suggested Dien Bien Phu as a suitable point to fight Giap, his

The strain shows on the faces of the French defenders of the isolated stronghold of Dien Bien Phu. Strewn on the ground are containers in which supplies were parachuted to them during the eight-week siege by the Vietminh.

deputy, General Cogny, warned him that it could become a 'meat-grinder' for French troops. Dismissing such negative thoughts Navarre insisted on concentrating a French force at Dien Bien Phu so as to force Giap to attack him. The result, Navarre assured Cogny, would be a crushing defeat for the Vietminh. But the logistical realities were rather different, as Navarre should have realized. A French army at Dien Bien Phu would be completely isolated and could only be supplied by air. The high ground surrounding the French base would be held by the Vietminh, who could be freely supplied from the north. Navarre's conception of what a guerrilla army could transport was again quite wrong. Using vast numbers of peasants with Peugeot bicycles, purchased from pre-war French shops and capable of carrying 500 pounds in weight, as well as every kind of wheeled, human or animal transport, Giap's soldiers supplied a force of 50,000 around Dien Bien Phu, outnumbering the French by four to one. Only as the cam-

paign developed did Navarre begin to realize that he had sent 13,000 Frenchmen into a trap from which they could not escape, even by air. The low clouds made flying dangerous and the entrenched Vietminh artillery made the French airstrip almost unusable. Even the tanks which Navarre had seen as a decisive weapon in the defence of the base were now seen to be useless. The terrain, which had looked so suitable on faulty French maps, turned out to be heavily covered in bush, which entangled the armoured vehicles and rendered them useless. The problem was compounded by the rains which, when they came, made the ground so soft that the tanks became bogged down and inoperable.

During November 1953 Giap moved 33 battalions of infantry and six regiments of artillery into the area of Dien Bien Phu. Showing remarkable expertise, the Vietminh gunners camouflaged and entrenched their guns to make them safe from air attack. Navarre's choice of commanders for the besieged base could hardly have been worse in the circumstances. The brilliant cavalryman Colonel Christian Marie Ferdinand de Castries was chosen because of his skill with the tanks, which the French were never able to use, while his deputy, the arrogant one-armed artillery commander Charles Piroth, airily told Navarre that 'no Vietminh cannon will be able to fire three rounds before being destroyed by my artillery'. But Piroth, humbled by his failure to match the Vietminh gunners, was later to pull the pin on a hand grenade and kill himself in his quarters.

Navarre had confidently told Cogny and de Castries that Giap would use the same 'human wave' assaults that he had so unwisely used in the Red River region in 1951. Now as then the French would massacre the Vietminh. But Giap had learned from his mistakes. And he was no longer paying so much attention to his Chinese advisers, who favoured such frontal assaults. Giap had decided slowly to strangle the French defenders of Dien Bien Phu. For three months he defied all French predictions by refusing to attack. But all this time his men were digging hundreds of miles of tunnels and trenches around the French base, as well as digging in their artillery and securing supply routes. Against the French force of 7,000 combat troops and 6,000 non-combatants, Giap assembled 50,000 assault troops and a further 20,000 men to guard his supply lines. Giap was aware that as weather conditions worsened, with low cloud and heavy rain, French dependence on air supplies would become a noose growing tighter by the day.

In the end the French professionals were undone by the professionalism of the man they despised, General Giap. The Vietminh used method and careful planning against French arrogance and flippancy. (The three artillery bases sited around the main compound at Dien Bien Phu – which were too far spread to support each other – were named after the commander's latest mistresses, Béatrice, Gabrielle and Isabelle.) Before the main base was attacked the two northern artillery strongpoints were taken, allowing Vietminh troops to close right in on the doomed defenders. With the military situation deteriorating the search for scapegoats began. Relations between Navarre and Cogny had reached a point where the two could not meet without the threat of physical violence. The only hope now was that the United States would agree to direct heavy air strikes on Giap's entrenched troops. A plan for using atomic bombs was even considered. But eventually the Americans decided that 'it was the wrong war at the wrong time'. Nothing now could save the doomed defenders. On 7th May 1954 the Vietminh overran the French defenders and Dien Bien Phu was taken, virtually ending the war in Indo-China.

With some justice Dien Bien Phu can be called 'Navarre's Folly', yet it must be remembered that his views were shared by a generation of European and American commanders. But they were to learn in Korea and later in Vietnam and Laos the folly of underestimating the skill and fortitude of guerrilla fighters.

CHAPTER 4: THE TACTICS OF DEFEAT

The Conservative

To ignore the lessons of past wars is to be doomed to repeat the mistakes of past disasters, yet conservatism in warfare has been at the root of many military blunders. The problem is not an easy one to solve. The military profession has often been a very conservative one, clinging to the traditional values of honour, duty, loyalty and courage, and believing that traditional methods such as the use of the bayonet or the cavalry charge can always be relied on whatever the circumstances. But the technology of war has advanced irrespective of the wishes of the generals, and those who have ignored this fact and followed the tactics of a past age have generally come unstuck (see also pp. 149–54).

'It must be accepted as a principle that the rifle, effective as it is, cannot replace the effect produced by the speed of the horse, the magnetism of the charge and the terror of cold steel.'

British Cavalry Training Manual, 1907

In spite of Britain's experience in the Second Boer War 1899-1902 and recent evidence from the Russo-Japanese War 1904-5, which showed the effectiveness of modern rifle fire, British military thinkers continued to believe in according the prime role in any encounter to the cavalry arm.

The medieval knight, archetype of tactical conservatism, received a rude shock at the beginning of the fourteenth century from his social inferior, the disregarded footsoldier. In a series of battles across the length and breadth of Europe, from Courtrai in 1302 to Bannockburn in 1314 (see p. 103) and Morgarten in 1315 (see p. 91), the message was the same: adapt or die. But the knights of France, pre-eminent in Europe in numbers and splendour, were not listening. And they were in for the greatest shock of all, at Crécy in 1346.

English lessons

Edward III's English army of 12,000 included as many as 8,000 archers, and had taken up an excellent defensive position along a ridge. The French army, led by King Philip VI in person, boasted a cross section of the chivalry of Europe, with German, Bohemian and Spanish knights adding their weight to the large French contingent. Three kings rode with Philip – Charles, King of the Romans, Jaime II of Mallorca, and the blind King of Bohemia, chained to a knight on either side to guide him into the fray. So many famous nobles were present that it must have been difficult to impose any kind of order on the huge cavalry formations, perhaps 12,000 strong. This was almost certainly the greatest

At the Battle of Crécy in 1346 the mounted French knights fatally underestimated their English opponents. Their frontal assaults against heavily defended positions were doomed to fail. The English longbows were capable of despatching ten arrows a minute and were effective up to 300 yards.

concentration of knighthood ever present in a medieval battle, easily capping the 2,000 knights at Bannockburn and 1,500 at Hattin. As in so many medieval battles, the knights were intensely proud, very arrogant and largely self-centred. It took a general of unusual ability to get the best out of them. Philip of France was not such a man; Edward III of England was.

It was late in the day when the French masses began to come in sight of the English position. At their head marched a force of 5,000 Genoese crossbowmen, highly efficient mercenaries who would have performed well under sensible direction. But the chauvinism of the French nobles made them contemptuous of these foreign footsoldiers, and although the Genoese had been marching for eighteen hours already and had wet bowstrings, they were hurried into action. The French pushed from the back and charged from the front as if passing down a narrow corridor. They had started the day in ten separate divisions, but everyone had closed up and so it was one heaving mass of men and horses that the English saw laid out below them.

Apparently Philip had ordered his men to make camp, but nobody took any notice.

With men calling out in many different languages, and common folk packing the surrounding roads and shouting, 'Kill! Kill!', it must have sounded like the Tower of Babel. The chronicler Froissart wrote that no one who had not been there could possibly imagine the confusion.

Whether King Philip wanted it or not the battle began when the Genoese crossbowmen advanced and fired a volley at the English. The volley fell short. They moved forward again, but before they could fire they were hit by a storm of English arrows. They had never faced anything like it before and panicked. As they tried to escape they were simply ridden down by the first line of the French knights commanded by the Duke of Alençon. This only served to throw the knights into more chaos. Before they could build up any momentum they were hit by the English arrow storm, which deservedly has the reputation of having been the most concentrated firepower from infantry before the nineteenth century. Probably 60,000 arrows were hitting the French knights every minute. Not surprisingly none of the first line reached the English position. Yet with the battlefield strewn with corpses of men and horses, the second line of knights rode straight into the maelstrom and were likewise shot down, among them the blind king of Bohemia.

Throughout the fighting, which continued during fifteen consecutive charges, the French knights made no attempt to turn the English flanks or attack the longbowmen. Their only aim seemed to be to engage the dismounted English knights in a battle of equals. Time and again head-on charges simply collided with the retreating remnants of the preceding one, like waves on a beach. None of the French commanders, whoever they were, were thinking tactically. They were following a pattern which their forebears had followed for centuries, through good times and bad. Here was one of the worst times. French casualties were prodigious, though only the figures for nobles and lords were accurately recorded: 1,542 killed, the cream of French chivalry. The like of this battle would not be seen again until Agincourt in 1415, when the French knights met much the same fate.

French lessons

At Formigny in 1450 and Castillon in 1453 the French were to have their revenge for Crécy. The English, unwilling to change a winning weapon in the longbow, succumbed to the killing power of the new French artillery. With the introduction of hand guns in the warfare of the sixteenth century, England's obsession with archery began to seem curiously archaic.

The other important role in the dual military revolution of the fourteenth century was played by the Swiss pikemen. They triumphed over the Austrian knights at Morgarten in 1315 (see p. 91), and held sway as mercenaries in Europe throughout the fifteenth century, but they too failed to adapt and so perished. Like the English they were too conservative to see that what once had been revolutionary was now outdated. At Bicocca in 1522 they were to suffer the same hard lesson that the English learned at Castillon. Believing themselves invincible and refusing to wait for their French allies to support them, 8,000 Swiss pikemen charged a powerful Spanish position made up of defensive hedges and ditches, lined with musketeers and pikemen. Undeterred, the Swiss leaped into a deep ditch in front of the Spanish position and tried to force their way up onto the other side. They were held down by the pikemen and shot to pieces by the gunners. Three thousand of them died in the ditch before they were forced to retreat.

General Braddock

Conservatism of thought can convince a commander that tactics he has used effectively in one situation will necessarily work in another, quite different one. This was the mistake made by General Edward Braddock in his fatal campaign in North America in 1755. Braddock was no fool, but he was a stickler for doing things by the book. His French adversaries had already found out that fighting conditions in the American forests were very different from those of Europe. They had adapted accordingly, changing their clumsy uniforms for Indian-style dress and moccasins. They had also learned to fight in the wild, taking cover behind trees and exploiting the conditions by laying ambushes. All of this was unknown to Braddock and his European soldiers. Before he had left England the Duke of Cumberland had reminded him that discipline was the fundamental military virtue and to forget this was to fly in the face of convention. Yet, as the British were slow to learn, flexibility was what was most needed in fighting irregular troops and Indians.

Marching from Fort Cumberland to besiege Fort Duquesne in June 1755, Braddock moved slowly, at a pace dictated by the men building the road on which they were advancing. The great dark forests of the Ohio region played on the fears of the ignorant English and Irish soldiers. They had heard wild stories of the cruelty of the Indians and they were psychologically half beaten even before they met the enemy. Meanwhile, Braddock continued on his leisurely way, flattening 'every molehill' and building a bridge over every creek. The French had no difficulty following the advance of the British column towards the Monongahela River and Braddock did nothing to conceal his progress. Although badly outnumbered, the French decided not to wait in their fort but to advance with their Indian allies and try to ambush Braddock. The British general, aware of this possibility, had sent out an advance force to spring any traps that might have been laid. But as he got closer to the fort, and the chance of ambush grew less, Braddock became careless. Perhaps the French were overawed and would retreat without a fight.

In fact, the French had decided on an ambush but had not been able to organize it before the British stumbled into them. Braddock's advance guard under Colonel Gage met a curious figure advancing towards them, dressed like an Indian but wearing a French officer's gorget at his neck. It was Captain Beaujeu, commanding the French and Indians. Beaujeu reacted quickly, sending his men scuttling into the trees. Gage formed his men up and fired three deafening volleys. But his men began to be picked off one by one by the unseen enemy. Cannons were now fired into the forest but did no apparent good. British flank guards, panicking at the sound of Indian war cries, ran from the cover of the trees and into the clearing, where they were shot down.

The whole British column now lapsed into confusion. With the vanguard retreating and the rearguard advancing, there developed a kind of rugby scrum in the middle of the clearing. Half the British casualties were caused by soldiers at the rear shooting down their own colleagues up front. Braddock, with great courage, charged around on his horse, bellowing orders. Having wrapped a white handkerchief round his hat, he presented a clear target for the hidden snipers. Yet though four horses were shot under him he seemed to lead a charmed life. One Virginian officer, showing more initiative than anyone else, ordered his men to take cover behind a fallen tree. But British officers, thinking these men were deserting, actually opened fire on them. When Braddock saw some of his redcoats sheltering behind trees he beat them back into the open with his sword. The British were forced to line up, as if on a parade ground, and fire volleys into the empty forests, suffering heavy casualties all the while. The massacre would have continued but for the fifth bullet missing Braddock's horse and shooting the general through a lung. With their commander mortally wounded, and 60 out of 86 officers already down, the remnants of the column turned and fled. George Washington, who had been with Braddock throughout and had tried to convince the stiff-backed British general to adapt more to the conditions, helped to organize the retreat. Braddock, with 45 years' experience of soldiering in Europe, had simply been unable to fight a battle to suit the circumstances. The dying general told Washington, 'We shall better know how to deal with them another time.' One would like to have agreed with him, but the next example makes it difficult.

Ticonderoga

Braddock was incompetent because he was too conservative. General James Abercromby was conservative and stupid as well. Only three years after the disaster on the Monongahela River, Abercromby was advancing to attack the French fortress at Ticonderoga with a strong force of 16,000 British and colonial troops. Opposing him was one of France's most able commanders, Luis Joseph, Marquis de Montcalm, with a mere 3,600 men. Montcalm ordered the trees in front of his fort to be cut down and used to build a powerful breastwork some nine feet high. He then covered the approaches to the fort with a mass of branches and lopped tree-tops. When Abercromby arrived with his troops he found that there were so many ways of attacking the fort that he was spoiled for choice. However, rejecting the idea of flank attacks on undefended parts of the fort as too easy, or artillery bombardment from hills overlooking the French positions as too obvious, or blasting the breastwork to pieces with his guns as too clumsy, he instead left his guns behind with the boats and ordered his infantry to charge through the thorn-bushes, hedges and branches laid down by the French. Perhaps he saw this as a thoroughly sporting way of evening up the odds. Having made his plans for a massacre, Abercromby retired from the battlefield and sat twiddling his thumbs in a saw mill two miles away.

His officers formed up their troops in lines of impeccable quality and ordered them to advance through the tangle of trees and branches. The French troops, who were firing through loopholes in the breastwork, were invisible to them. Highlanders, swords in hand, tried to clamber up the French defences only to be bayoneted or knocked down. The ridiculous struggle continued for five hours, with Abercromby distantly ordering 'Advance and attack.' No attempt was made to try anything other than a frontal assault. In all the British suffered over 2,000 casualties because of the obstinate stupidity of their commander. His belief in the superiority of British troops, advancing in steady order, was totally misplaced. Perhaps General Braddock turned slightly in his grave.

New Orleans

Wellington's Peninsular veterans rightly have a high military reputation as some of the toughest and steadiest fighters of their time. Yet under the wrong commander even these élite troops could fail. The story of their use at New Orleans in 1815 by Sir Edward Pakenham, Wellington's brother-in-law, is further confirmation of the falsity of the dying Braddock's words. Pakenham had fought alongside Wellington through some of his greatest battles, but had learned about as much about commanding an army as the Iron Duke's boots. Even Wellington himself had referred to Pakenham as not exactly 'the brightest genius', and this in an army boasting such worthies as Sir William Erskine and 'Black Jack' Slade (see p. 43).

Many of the Peninsular War veterans under Pakenham's command believed they were being taken back to Britain for discharge after seven years' service under Wellington. Their disappointment at finding themselves fighting in the New World must have been a factor in what was to happen when they next came under fire. The fact that the British artillery was short of shot so that men of the 7th Fusiliers and the 43rd Foot had to stagger to the front burdened by cannon balls in their knapsacks can hardly have improved anybody's temper.

Although the war between Britain and the United States had been officially ended by the Treaty of Ghent in 1814, nobody seemed to have told American General Andrew Jackson or Pakenham, and the most decisive battle of the war was still to be fought. Jackson, with some 5,000 troops, was defending a position near New Orleans. His troops were not of the first quality, some being regulars, while others were militiamen, civilian volunteers, freebooters and pirates, and even some freed slaves. In hand-to-hand fight-

Scenes such as this are more popular with artists than they ever were with soldiers. The bayonet has been the most overrated and most underused weapon in military history. Percentage rates for casualties caused by the bayonet in wars since the eighteenth century are negligible.

ing they could not have stood against Pakenham's British redcoats. But Jackson had no intention of asking them to do that. By choosing a strong defensive position, anchored on the left by a swamp and on the right by a river, Jackson was inviting the British to assault his front of less than a thousand yards. With 5,000 muskets covering such a front a head-on attack would be suicidal.

But Pakenham believed that steady troops could take Jackson's position by the bayonet (it is usually fatal when a commander reaches this conclusion). As the redcoats walked forward, as impressive a sight as at Ticonderoga, they were hit by a devastating fire. The rout of his first line did not deter Pakenham. He concluded that if he were to lead the second line in person the result would be different. It was not and he was killed. British casualties of 2,100 had to be balanced against American losses of 8 killed and 13 wounded. For the Americans it was not a battle, it was more like live shooting practice. For the British it was a lesson which they obstinately refused to learn and so suffered the consequences.

Roughing It

Field Marshal Haig's decision to launch an offensive in Flanders during the summer and autumn of 1917 was not in itself a military blunder. There were sound strategic reasons for the decision, but Haig failed to give enough consideration to the likely effects on the terrain, low-lying and prone to waterlogging, and the forecast from the Meteorological Office of rainy weather. To compound these errors, Haig introduced a factor which guaranteed that the terrain would be hostile. He proposed to batter the German positions with the biggest artillery barrage of the war so far, even though he had been warned that the drainage system of Flanders was based on a delicate balance between man's skill and the forces of nature. Before Haig made his decision to launch the battle of Passchendaele he was warned by many people, including the authoritative Belgian government department responsible, that if he bombarded the land around Ypres it would become a swamp. It had only been reclaimed from marshland by centuries of hard labour. It could be lost in a few days of human folly.

> *'We must never forget that we are dealing with men of flesh and blood and nerves.'*
> **Major-General Douglas Haig, 1903**
>
> Haig must have forgotten his own words in 1916 on the Somme and in 1917 at Passchendaele.

Haig refused to listen and as a result was to blame for what happened next. On 22 July 1917, 3,091 British guns, including 1,000 heavy mortars and howitzers, began to pound the whole area with 4.25 million shells. Aerial photographs taken afterwards show a scene of unbelievable desolation, as if an area of the moon had been hit by a concentrated meteor storm. On every square yard of the ground nearly five tons of high explosive had fallen. The result was that the drainage system collapsed and the area was turned into a swamp by heavy rain. Not content with this, Haig then ordered hundreds

Canadian troops at Passchendaele in 1917. Haig's preliminary bombardment of German positions around Ypres turned the area into a swamp, and the British offensive became bogged down in a nightmare of mud.

of thousands of men to fight on foot in mud so deep that 90 men a month drowned in it. Assault after assault was ordered against German positions, many of which clung to what ridges remained, and the Germans were able to fire from firm footholds on British troops wading forward waist-deep through mud. From the comfort of their château headquarters miles from the front, Haig and his staff moved flags on wall maps and never knew what the conditions were like because they did not care to look for themselves. Such generalship defies description.

The Mud March

Mud is no friend to any general. Yet compared to the horrors of Passchendaele, General Ambrose Burnside's 'Mud March' after the battle of Fredericksburg in 1862 has the refreshing air of farce always associated with this commander. After the slaughter at Marye's Heights (see p. 116), Burnside's men settled down behind the Rappahannock River in Virginia to spend a dismal, wet Christmas in the open. A period of dry weather at the turn of the year persuaded Burnside to look for a new crossing point. On 20 January 1863 he got his army moving along the river, only for the heavens to open and turn the whole area into a swamp. As the Union soldiers waded through the mud, the guns sank never to be seen again and the baggage animals had to be hauled out of the mud with ropes. The march averaged a mile a day. Meanwhile, on the other side of the river, General Lee fortified every crossing point. The Confederate soldiers laughed at the Union soldiers' discomfort and held up signs saying 'Richmond, this way', or yelled ironic offers of support, 'Hey, Yankees, shall we come over and help you build a bridge?' Burnside was furious. When he reached United States Ford he determined to launch another Fredericksburg-type crossing, but his generals stopped him. It was the end of the line, at least for the moment, for Ambrose E. Burnside, who had destroyed the morale of his men. Under pressure from Burnside's divisional commanders the thoughtful Lincoln removed old 'Side-burns' from command of the Union Army.

Morgarten

Over six hundred years earlier, in the mountains of Switzerland, an Austrian commander was finding things just as hard. The Habsburg duke Leopold II, with a strong force of 2,000 knights and 7,000 infantry, was trying to suppress a rising among the hillfolk of the Swiss cantons. He was to come to grief near Morgarten in 1315 as a result of faulty reconnaissance and trying to fight a battle in surroundings unsuitable for his heavily

The Swiss handed out a harsh tactical lesson to Leopold II of Austria in mountainous terrain at Morgarten in 1315. On steep slopes unsuitable for heavily armoured mounted knights, the Austrians were slaughtered by agile Swiss footsoldiers.

armoured knights. Using his knights to lead the army up the slopes of a mountain was a particularly grave blunder. As the track narrowed his men had on their right the sheer mountainside, while to their left was the drop into the Lake of Egri – an unenviable position for horsemen facing the agile footsoldiers of the Swiss mountains. Leopold found that the road ahead had been blocked by a wall of stones and boulders, behind which was a small guard of Swiss soldiers. The bulk of the Swiss force, some 2,000 strong, was hiding in the wooded hillside above the winding path, on which the Austrians were advancing. As the Austrian vanguard tried to engage the Swiss defenders at the barricade, the rest of the army kept pushing forward, creating confusion in the narrow space. Suddenly, boulders and tree trunks were rolled down from the slopes above, tipping many Austrians, horses and riders over the cliff edge and into the water below. To the sound of horns the halberd men who comprised the bulk of the Swiss force rushed down the slopes to hit the Austrians in the flank. Hemmed in on all sides the helpless knights were slaughtered or driven screaming over the edge onto the rocks below or into the deep blue water of Lake Egri. For the knights at the head of the army there was no escape and the majority of them were slain. Leopold, whose poor generalship had led to the disaster, was hurried back to safety. This triumph was but the beginning for the Swiss pike and halberd men, who were to become the foremost infantry of their day, and to dominate European battlefields for nearly two centuries.

Lake Trasimene

Hannibal's ambush of the Roman Consul Flaminius at Lake Trasimene in central Italy in 217 BC bears some similarities to the Swiss victory above. Yet in both cases the victors were dependent on their opponents' lack of tactical skill – notably their poor reconnaissance. Hannibal planned to ambush the pursuing Roman army along the shore of the lake by blocking the exit from the valley. He occupied the wooded hills overlooking the lake with his Iberian and African troops and waited. The heavy mist that rose over the water and shrouded his army in the hills hardly excuses Flaminius's failure to locate so large a force. Instead, he hurried his troops into the valley and along the coastal paths, unaware that Hannibal was waiting for him. The mist was so thick that the Romans could scarcely see more than a few feet ahead and yet with the chances of an ambush so strong they appear to have taken no precautions. When Hannibal gave the signal to attack his men poured down the hillsides and drove the Romans to the edge of the lake, where many of them were drowned. In all nearly four legions were destroyed.

The fog of war

If Flaminius fell victim to a very real 'fog of war' at least it did not cause the confusion that struck the Lancastrian army at the battle of Barnet in 1471. Here a thick dawn mist caused the Earl of Warwick's men to mistake the silver star with streamers that was the device of their ally, the Earl of Oxford, for the 'sun of York' – the emblem of Edward IV, the Yorkist leader. Warwick's men immediately fell on their allies and in the confusion the battle was lost and Warwick killed.

At the battle of Philiphaugh in 1645 during the English Civil War the failure of the

Marquess of Montrose – otherwise a brilliant and dashing commander – to anticipate the effects of an early morning fog led to defeat. He had camped near Selkirk on very low ground, apparently to save the trouble of fetching water from the nearby river. However, the month being September, a thick dawn mist enshrouded his whole camp, and the Parliamentarian army of David Leslie, coming from the hills, was able to descend on him and take him unawares. The poor visibility made it impossible for Montrose to form up his army to receive the attack and his camp was simply overrun.

The Winter War

During the so-called 'Winter War' between Finland and the Soviet Union in 1940, the Soviets blundered by disregarding the nature of the land in which the war would be fought as well as underestimating their enemy. Their belief in modern technology – the tank, the plane and motorized troops – and their contempt for the simplicity of Finnish methods, led to humiliating and costly defeats. The Soviets tried to use the 'blitzkrieg' tactics that the Germans had used against Poland in 1939, but forgot that they were faced by the frozen forests of Finland rather than the wide plains of Poland. Here the individual Finnish fighter, equipped with Molotov cocktails and skis, was more than a match for the Soviet forces, fighting on foot or in motorized vehicles which simply froze in the arctic conditions. Most of the modern Soviet weapons malfunctioned during the winter campaigning and had to be redesigned.

The Soviets found themselves fighting a war they were ill-prepared to fight. The cold weather rendered weapons, tanks and trucks unusable. Food supplies had to be far heavier than normal to keep the men alive and this placed an impossible burden on the transport system. Medical services were unable to cope with the situation and most wounded Soviets froze to death in minutes. Temperatures were so low that one Finn, shot six times through the chest, was able to keep walking to the medical station and receive attention because he bled so slowly. In contrast with the many layers of clothing worn by the Finns, the Soviet tunics were almost useless in the extreme cold. Many Soviet soldiers even lacked tents and had to dig makeshift shelters in the icy ground to keep themselves alive. Morale in the Soviet forces plummeted with the temperature. The presence of political commissars who overruled their officers meant that no unit was allowed to surrender, however hopeless their position. Soldiers were told that if they gave up their families in Russia would suffer, while they themselves would be tortured by the barbarous Finns. Many soldiers, including officers, committed suicide rather than surrender.

Twenty Soviet divisions spearheaded the assault on the northern wilderness of Finland. The infantry had been marching for over 200 miles from the last railhead at Murmansk, and thousands of men had already dropped out with frostbite. Lacking proper tents, most of the men crouched around blazing fires all night in an attempt to stay alive. Soviet ignorance of what to expect in Finland was staggering. They anticipated being able to use their tanks and trucks on Finnish roads, but in fact encountered snow drifts ten feet deep. In some places, by a quirk of nature, the ground was warmed by the snow above and was actually swampy as the Soviets tried to walk on it. Sinking through the snow into a hideous slime during the day, the men were then frozen solid by the freeze which set in during the night. Vehicles could only pass along the roads if pre-

In the appalling conditions of the Soviet–Finnish Winter War of 1939–40 reliable – and resilient –
equipment was crucial. Here Swedish volunteers anxiously assess the quality of their skis. Excellent
Finnish equipment enabled the outnumbered Finns to inflict heavy losses on their Soviet opponents.

ceded by groups of infantry 200 strong and wearing heavy boots, who flattened the
snow. When one considers that a Soviet division could stretch some 20–25 miles, and
that its progress was limited by the snail's pace of men trudging through heavy snow,
one gets an idea of the scale of the miscalculation the Soviets had made in attacking Finland in winter and with heavy armour.

Soviet losses in the war stagger belief. According to Nikita Khrushchev over a million
Soviet soldiers died against Finnish losses of 25,000 dead. In addition, Soviet material
losses were heavy: 1,000 aeroplanes, 2,300 tanks and armoured vehicles and vast
quantities of munitions, transport and horses. The poor Soviet performance convinced
Hitler that he had nothing to fear in invading Russia in 1941. However, the Russians
learned a lot from the harsh lesson of war in Finland and replaced lost equipment with
far better material, notably the fine T-34 tank – a war winning weapon (see p. 175).

A Poor Lookout . . .

'In war,' said Napoleon, 'the simplest operations are the best and the secret of success
lies in simple manoeuvres and in taking measures to ensure against surprise.' Gathering
information about an enemy and reconnoitring his position has been a basic strategic
requirement throughout military history. Every plan of operations, whether prepared

by an Assyrian king, a Roman consul, a Byzantine general, or a modern commander, needs to be based on accurate intelligence reports. Moreover, forward reconnaissance will also prevent the enemy surprising him and improve his chances of obtaining surprise himself. Any commander who fails to reconnoitre his enemy's position is likely to suffer a self-inflicted wound, often a fatal one.

Caught napping

By 1066 the Norwegian king Harald Hardrada had won himself an unrivalled reputation as a skilled commander and as the greatest Viking warrior then living. A long career in the Byzantine Varangian Guard had given him a breadth of experience against different foes that no one could match. And yet the one serious mistake of his career cost him both the destruction of his army and his life. After his successful struggle against the northern English *fyrd* (militia) at Fulford in 1066, he let his guard drop for a moment, allowing his rival, Harold Godwinson, to take him by surprise at Stamford Bridge four days later. Confident that Harold's southern army was still on the south coast awaiting the arrival of the Normans, Hardrada had encamped his army on meadows either side of the River Derwent, not far from York. So unseasonably hot was the weather that many of the Vikings had stripped off their armour and were bathing or idling in the summer sun. But Godwinson had force-marched his men from London and, encountering no Viking pickets, swept across the meadows taking Hardrada by surprise. Even the approaching clouds of dust, followed by the sun glinting on spear points and armour, did not stir the Vikings, who thought it was merely a detachment of Norwegians approaching from the fleet's anchorage. The Vikings on the west bank of the Derwent were soon overrun, few even having weapons or being able to don their armour. In spite of the heroic defence of the bridge by a single Viking Berserker, the Saxons crossed to the east bank, where Hardrada had managed to get his men organized into a shield wall. Here the fighting was long and bloody, but the result was never really in doubt. His losses early in the fight gave Hardrada little chance and he was killed along with most of his army, all victims of an elementary mistake on the part of an otherwise great commander.

Unpleasant surprises . . .

Faulty reconnaissance has been at the root of several famous military disasters in modern times. Even Napoleon was not immune. His failure to locate the Austrian troops of Archduke Charles at Aspern-Essling in 1809 contributed to his first military setback. During both the Franco-Austrian War of 1859 and the Franco-Prussian war of 1870 the standard of reconnaissance was so poor that 'encounter' battles, in which regiments and divisions just blundered into each other without proper instructions from their commanders, were common.

In the early years of the American Republic, two prime bunglers in the shapes of Generals Harmar and St. Clair suffered entirely avoidable defeats at the hands of the Indians. In June 1790 Harmar led a motley force of nearly 1,500 men, only one in five of whom were regulars, into Indian country. Instead of concentrating his strength, he broke his force up into 'penny packets', who proceeded to spread out in unruly fashion looking for plunder. Having abandoned the normal precautions, two of his detachments were ambushed and wiped out, forcing Harmar to retreat hurriedly back to Cincinnati. General St. Clair set off in September 1791 with some 2,000 men, to see if he could do any better. But he had completely misread the reasons for Harmar's failure and believed that he must not split up his force in the slightest way. He decided, therefore, not even to send out scouts, and his column advanced blindly into Indian country. On 3 November, near the headwaters of the Wabash river, 1,000 Indians crept up unnoticed on his camp. In terms of health St. Clair was thoroughly unfit to be leading the column anyway, being bed-ridden through rheumatism, asthma and colic. Just before dawn the Indians fell on the camp, killing 637 men and wounding many others. Somehow St. Clair and half of his force survived to return to Cincinnati. But the lesson was clear. Fighting Indians with amateurs was a dangerous business and without reconnaissance it was plain suicide.

In colonial wars it was always necessary to maintain an effective scouting system against technologically primitive but fast-moving opponents. At Isandhlwana in 1879 Lord Chelmsford's faulty reconnaissance allowed a Zulu army of 20,000 to remain undetected until it burst upon an unprepared camp (see p. 132). The appalling failure of the Italian commander Baratieri to locate the numerically superior Ethiopian army at Adowa in 1896 allowed his four separate columns to be surprised and overwhelmed one after the other (see p. 172). Colonel Armstrong Custer broke almost every principle of military tactics and strategy at the battle of the Little Big Horn and launched an ill-prepared assault against a vastly superior enemy and was wiped out with his whole command (see p. 21).

Dummy soldiers and 'guns' made of painted logs were a feature of the early part of the American Civil War. They deceived General George McClellan at Munson's Hill in 1862 and supported his belief that he was outnumbered by Lee's Confederates.

At the beginning of the American Civil War, Union general Robert C. Shenck was leading a patrol in Virginia, without scouts or advanced guards. He blundered into a Confederate ambush, but rather than admit his mistake he claimed that he had been fired on by 'masked batteries' (concealed artillery). In fact the Confederates only had two field guns in addition to their infantry. However, two cannon would not justify Shenck's defeat, so he found it necessary for the sake of his own reputation to invent hidden batteries of guns which had, rather unfairly, brought about his downfall. Shenck's concocted excuse made him the unwitting instigator of a powerful myth. Like the stories of the Loch Ness Monster or little green men from Mars, the story of 'masked batteries' snowballed until it had entered the soul of every Northern commander. Soon every Union column moved in fear of what might strike them unawares. At the first battle of Bull Run General McDowell chose the watchword 'Caution' as the order of the day. Without knowing it, and through a simple reconnaissance failure by the enemy, the Confederates had gained a psychological advantage over the North which they were not to surrender until after Gettysburg, two years later.

Crossed Lines . . .

Communications between a commander and his subordinates during a battle are so important that one would assume that they would be given the utmost priority. There should be no room for misunderstanding. Yet it was a breakdown in communications which caused the defeat of the Union army under General Rosecrans at the battle of Chickamauga in the American Civil War.

'Like a duck hit on the head'

On 20 September 1863 the troops of Confederate general Braxton Bragg and Union general William S. Rosecrans were locked in confused fighting. The first day of the battle had been indecisive and a nervous Rosecrans feared that his left flank, under George Thomas, was in danger of giving way. To bolster Thomas's wing he decided to shift extra troops from his centre. Forgetting to inform General McCook, who was commanding the centre, Rosecrans gave a direct order to a divisional commander, General Wood, to shift his men to the left. In the heat of the moment, Rosecrans made the mistake of thinking Wood's division was in reserve, behind McCook's front line. In fact, Wood's division was in the front line and to move it would be to create a huge hole. Not realizing this, Rosecrans dictated a brief note to his ADC, Major Bond, and told him to give it to Wood. The note is very short and very misleading.

> To General Wood,
> The general commanding directs that you close up on Reynolds as fast as possible and support him,
> Bond.

When he received the note Wood was astounded. Between his own and Reynolds's divisions there was another division under General Brannan, which covered a quarter of a mile of front. Was he supposed to pull his men out of the front line and march round the back of Brannan's men in order to reach Reynolds? Since this is what the order said this is what Wood felt obliged to do. The result was quite predictable; some four hundred yards of the Union front line disappeared, directly opposite Confederate general Longstreet's men.

When excited officers told Longstreet what had happened he acted with great speed, ordering his own General Stewart's division to charge through the gap, and supported him with six more divisions, totalling 30,000 men. The Confederates fanned out on all sides, taking the Union troops in the rear and breaking the Union army in half. Rosecrans's right flank collapsed and was driven off the field. His artillery and baggage train were captured and Rosecrans himself fled all the way back to Chattanooga. Whether the fault was his or Major Bond's is less important than the fact that Wood was obliged to obey an order which cost his commander the battle. The fact that George Thomas held the left side of the Union army together and earned the nickname the 'Rock of Chickamauga' did nothing to save Rosecrans's career. As a commander he had lost his nerve and according to President Lincoln acted, 'confused and stunned like a duck hit on the head'.

A note of discord

Rivalling the drama of Rosecrans's confusing letter is the even more famous note sent by Lord Raglan to his cavalry commander, Lord Lucan, at the battle of Balaclava in 1854. The message from Raglan was the direct cause of the Charge of the Light Brigade – perhaps the most notorious military blunder in history.

The Russian field commander in the Crimea, Prince Menshikov, had direct orders from St. Petersburg to relieve pressure on Sebastopol by attacking the British base at Balaclava. On 25 October he attacked the port in strength. The British resisted in inimitable style, turning defeat into victory, setbacks into triumphs, and brought to the science of warfare the clumsy antics of the public schoolboy. The 93rd Highlanders created one legend, that of 'the thin red line tipped with steel'. The Heavy Brigade made one of the greatest and most effective cavalry charges of the nineteenth century, which was immediately forgotten. The Light Brigade made one of the most ill-advised charges in all military history and was remembered evermore. It was a curious battle.

The disparity between what Lord Raglan, British commander-in-chief, meant in his message to Lord Lucan and what Lucan understood by it was partly a reflection of the confusion in the minds of the men involved. It was also a product of the different positions they had taken up and their dissimilar fields of vision. The first message read, 'Cavalry to advance and take advantage of any opportunity to recover the Heights. They will be supported by the infantry, which have been ordered to advance on two fronts.' Lucan was uneasy on receiving this note and did nothing for 45 minutes. Did Raglan mean that he should wait for the infantry to arrive before he did anything? If so, where were the infantry?

From his vantage point Lord Raglan could see that his orders were not being followed. General Scarlett's Heavy Brigade had thrown the enemy into disorder. Why had the Light Brigade not reinforced that charge and swept the Russians from their redoubts? What on earth was Lucan doing just sitting waiting? As everybody waited, the Russians began to drag away some British guns they had captured. This was too much for

'"Forward the Light Brigade!" / Was there a man dismay'd? / Not tho' the soldier knew / Someone had blundered: / Their's not to make reply, / Their's not to reason why, / Their's but to do and die.'
Fired on by guns from both flanks and front, the men of the Light Brigade ride into 'the valley of Death' at Balaclava in 1854.

Lord Raglan, who dictated another order to General Airey, who scribbled the words on a piece of paper before giving it to an ADC, Captain Nolan. It is doubtful if this message was any clearer than the last one. 'Lord Raglan wishes the cavalry to advance rapidly to the front – follow the enemy and try to prevent the enemy carrying away the guns. Troop Horse Artillery may accompany. French cavalry is on your left. Immediate. Airey.'

Captain Nolan, renowned rider and hothead, spurred his horse down the hill. As he departed he heard Raglan's final words to him, 'Tell Lord Lucan the cavalry is to attack immediately.' But Lucan was still unsure, even when Nolan insisted he should attack at once. What was he supposed to attack? At ground level the situation was less clear to him than it was to Raglan on the heights. What guns was Raglan referring to? At this Nolan gestured furiously, 'There is your enemy! There are your guns!' But Lucan still could not see any Russians towing away guns. Why he did not send anyone to a high point to identify his targets is inexplicable and it was this failure that was to lead directly to disaster. The only guns he could actually see were sited at the far end of the North Valley, where masses of Russian cavalry were concentrated. He assumed, therefore, that these were the guns that Raglan meant. Pointing down the valley towards the Russian guns, Lucan ordered Lord Cardigan to lead the Light Brigade to the attack. Both Lucan and Cardigan knew that it was a suicidal order, but felt they had no alternative but to obey. Resplendent in gold braid at the head of his 'cherry bums', Lord Cardigan led forward his brigade of 673 men. Nolan had insisted on accompanying the Light Brigade and at once took the lead, riding ahead of Cardigan to his lordship's fury. In fact, it was Nolan who died first as his chest was ripped open by shell splinters. General Bosquet, watching from the heights, muttered his immortal words 'C'est magnifique mais ce n'est pas la guerre', which the British have proudly assumed was a compliment, omitting the second part of the Frenchman's observation which sets the whole in perspective – 'C'est de la folie.'

There were post mortems in abundance. Raglan's verdict was clear from his remark to Lucan, 'You have lost the Light Brigade!' Perhaps he was right, but in the heat of battle there is little time for interpreting an apparently baffling order from a commander. A subordinate officer must simply obey the order he believes he has been given. If the consequences are disastrous then the blame belongs with the commander, whose orders in this case were not so crystal clear as to admit of no second interpretation.

'We would rather have a classically educated boy than one who has given his mind very much up to Electricity and Physics and those kind of subjects . . . Power of command and habits of leadership are not learned in the laboratory . . . Our great point is character; we care more about that than subjects.'

Lieutenant-Colonel Murray, Assistant Commandant at Woolwich, 1902

The British military establishment was afraid of 'clever soldiers' or 'specialists' because they might challenge the supremacy of the gentlemen officers who had led British armies for centuries. A classical education was the hallmark of the public school boy. Science and engineering were more the field of the middle and lower classes, who made good NCOs but were considered to lack the character and moral requirements for leadership.

THE BATTLE OF BENEVENTUM
(275 BC)

At the battle of Malplaquet in 1709, the Duke of Marlborough and Prince Eugene suffered some 24,000 casualties as against the 12,000 suffered by the French and yet counted themselves victorious. It was a classic example of a 'Pyrrhic' victory. But who was Pyrrhus and why is he remembered for winning battles that others would count as defeats? Pyrrhus was king of the Greek state of Epirus in the third century BC. He was the greatest commander of the century after Alexander the Great and combined the Macedonian phalanx formation with the use of elephants in battle, as the successors

of Alexander had learned to do. In 280 BC he was called to the aid of the Greek city-states of southern Italy, which were facing conquest by the Romans. In fighting these Roman barbarians, as Pyrrhus obviously regarded them, he was matching the classic Greek method of fighting against the emerging legionary system of the early Roman Republic. It was bound to be a fascinating struggle.

In a series of battles over the next five years Pyrrhus won victory after victory over the Romans. But each was harder than the last and it became obvious to him that he was facing a special kind of barbarian, who learned from each setback and fought harder next time. After the battle of Asculum Pyrrhus commented, 'Another such victory and we shall be ruined.' He could not replace the veteran troops he was losing in this attritional warfare, whereas the Romans seemed to have

The Battle of Beneventum in 275 BC saw a clash between the fighting methods of the ancient Greeks and the emerging Roman Republic. Pyrrhus's use of elephants had spread panic among the Romans in earlier encounters, but at Beneventum they opened ranks to let the elephants pass harmlessly through to the rear where they were jabbed in the trunks by waiting spearmen.

limitless manpower on which to draw. Continuing the war would inevitably lead to disaster. Yet when disaster came it was the product not just of Roman numbers and determination, but of mistakes which Pyrrhus was to live to regret.

Firstly, he placed too much reliance on his elephants, which were an unpredictable weapon at the best of times. When the Romans first encountered the great beasts they were defenceless through terror, but in later battles they developed tactics for dealing with them. As a weapon elephants could be two-edged. If they became panicky they could trample their own side, as had happened to the Persians during Alexander's campaigns. Used sparingly the elephant was an excellent weapon, but if used too often their shock value was lost and they could even become self-defeating.

Secondly, Pyrrhus planned to make use of a night march, one of the hardest of all military manoeuvres, and rarely used successfully in the ancient world. If such a march were to go wrong, losses could be heavy.

At Beneventum in 275 BC, Pyrrhus was desperate to prevent the joining together of two Roman armies, both outnumbering his own. He decided to deal with each army in turn, beginning with that of the Consul Manius Cunius. He hoped that the night march would enable his troops to fall on the Roman camp from the rear at the first light of dawn. He selected an élite corps of warriors to make the march, through a forest and up a steep slope to command a ridge. Before ordering the troops to set off, Pyrrhus had an unpleasant dream in which all his teeth fell out and his mouth was filled with blood. Being superstitious, Pyrrhus believed the dream was an augury of disaster and tried to cancel the orders for the night march. But his generals told him that it was much too late to alter the plan and the march went ahead.

The troops set off into the darkness, taking the best elephants with them, and guided by men with

The victory over the Romans of the German chieftain Arminius in the Teutoburger Wald in AD 9
was the result of a successful ambush. Trapped on narrow forest paths, three Roman legions under
Quintilius Varus were unable to deploy and succumbed to the furious assaults of the Germans.

torches, for the night was intensely dark. Unfortunately they had miscalculated the distance involved and all the torches went out before the wood had been passed. The result was that the Epirotes were lost in the forest and only found their way to the ridge once it was light enough to do so. But they were easily spotted by the Romans, who attacked them and soon wiped them out. Manius Cunius now turned on Pyrrhus's main army. During the battle that followed, Pyrrhus's phalanx was wrecked by a young elephant which, maddened by spear points, rushed about the field seeking its mother. Pyrrhus had tasted defeat for the first time against the Romans, and withdrew from Italy, leaving it, as he said, to be a battleground for Rome and Carthage.

THE BATTLE OF TEUTOBURGER WALD (AD 9)

The commander who fails to carry out effective reconnaissance and permits the enemy to surprise him, particularly on a march through hostile country, is guilty of self-imposed injury. According to the Roman historian Velleius Paterculus, 'a sense of security' was the principal reason why Publius Quintilius Varus allowed himself and his three legions to be ambushed in the Teutoburger Forest by the Cherusci under Arminius in AD 9. Varus was an arrogant and rather stupid man – a mixture that generally makes a military disaster inevitable.

Varus had previously been Governor of Syria and 'was more accustomed to the leisure of the camp than to actual service in war'. He was related to the Emperor Augustus and was something of a court appointment, lacking the drive necessary to govern the German tribes. He made the cardinal error of treating the Germans as if they were slaves of Rome and as passive in character as the more sophisticated and Romanized Syrians. He relaxed the discipline of the legions, neglected training and manoeuvres.

In September of the year 9, the calm was broken by reports of minor tribal risings. On his way to winter camp, Varus decided to pass through the troubled area – between the rivers Ems and Weser in northwest Germany – and restore order. Regarding the rising as no more than a localized disturbance he took no precautions to place his troops on a war footing, and allowed families to travel with the troops. In fact, Varus was marching into a trap planned by Arminius, son of the leader of the Cherusci, a 'frantic spirit' who nursed a fanatical hatred of the Romans.

Arminius knew that no one succumbs to an ambush as easily as the commander who feels completely secure, and he engineered the minor rising so that Varus would have to investigate it by marching first through the friendly territory of the Cherusci. Once the Roman columns had penetrated deep into the forest they would be attacked by Arminius' German hordes. The plan for the ambush was revealed to Varus by Arminius' uncle, who hated his young nephew, but the Roman commander rejected the claim as a product of interfamily rivalry and neglected to examine the facts.

In the early autumn, Varus set out with three legions and a long column of baggage wagons and camp followers. Surprisingly he was accompanied by Arminius and an escort of Cherusci, probably to settle any fears the Romans may have felt about his loyalty. All went well until suddenly, deep in the Teutoburger Forest, Arminius and his Germans disappeared. Then news came from the head of the column that small groups of soldiers had been attacked. Amongst the swamps and treacherous paths of the forest, Varus was now facing an emergency. To add to his problems a sudden storm turned all the paths into quagmires and the high wind broke off branches which crashed down startlingly among the troops. Probably the darkness and atmosphere of gloom served to depress the Romans even more.

Suddenly the Cherusci and their allies struck along nearly the whole length of the column, hurling javelins into the Roman ranks. With the legionaries unable to deploy, the fight became a chaotic hand-to-hand struggle in which superior Roman tactics could not be used. Having repulsed the first German waves, Varus built a fortified camp and burned his wagons since they were now useless in the soggy conditions. The Romans battled on against constant attacks the next day, but in driving rain their bowmen found their weapons almost useless and the legionary shields becoming sodden and too heavy to carry. Arminius, scenting victory, closed in for the kill. Afraid of being taken

alive Varus and his senior officers killed themselves, while a few contingents of cavalry managed to hack their way free. The vast mass of the Roman column – soldiers, women and children – were either slaughtered in the forest or taken alive to be sacrificed to the German Gods. The horrifying remains of these sacrifices – later found by Germanicus, grandson of the Emperor Augustus – turned the stomachs of even the toughest of his Roman soldiers.

When Augustus heard the news of the disaster he panicked, expecting at any moment to see Germans marching through the Forum. Later Romans, more accustomed to disasters, would take such defeats more rationally. However, it was not long before Germanicus was back on the Rhine punishing the tribes for what they had done. But things had changed. Through the careless stupidity of Quintilius Varus, Rome's prestige north of the Rhine could never be restored completely.

THE BATTLE OF BANNOCKBURN (1314)

Edward II has an unenviable reputation among English kings. Yet on two counts at least he stood head and shoulders above any other medieval monarch: he died the most appalling death of any of them and suffered the most disastrous English defeat since King Harold at Hastings. The victory of Robert the Bruce at Bannockburn undid all the work of Edward's great father Edward I, the 'Hammer of the Scots', and secured independence for Scotland.

Yet it could all have been so different. Edward II had at his disposal overwhelming superiority in numbers and tactical options. Having brought a powerful English army into Scotland it was his dubious achievement to get it stranded in a position where it could not manoeuvre, leave his potentially 'battle-winning' archers underused at the rear, then to compound the folly by quitting the field before the fight was over, causing the rest of the army to give up and flee. It was a ludicrous display of how not to lead, control and motivate a medieval army.

The apparent purpose of Edward's expedition in

The Scottish victory at Bannockburn was won by a mixture of Scottish skill and English blunders.
Edward II of England cancelled out his superiority in numbers and tactical options by relegating his
potentially battle-winning archers to the role of passive watchers.

1314 was to relieve the Scottish siege of Stirling Castle. But his real reason was to impose his rule on Scotland once and for all. Yet his unpopularity with his own nobles meant that only three of his earls supported him, and such powerful men as the Duke of Lancaster and the Earls of Warwick, Arundel, Surrey, Oxford and Norfolk refused to send more than the bare minimum of mounted troops. More than the loss of their knights, Edward was to miss the military advice they could have given him. Nevertheless, the English barons were well represented, and among these were many experienced warriors, including Aymer de Valence, who had fought Bruce twice in pitched battles and Sir Giles d'Argentan, who bore the unusual title of 'third best knight of his day'. Edward's total force probably included over 2,000 knights, 15,000 infantry and large numbers of archers.

Against Edward's numbers, high indeed for a period when logistics were difficult and primitive, Bruce could field just 6,000 footsoldiers and perhaps 500 horsemen. However, the English levies were of doubtful quality compared to Bruce's pikemen, who were battle-hardened, well trained and fighting under trusted commanders for a cause they loved. The real test would come in the struggle between the Scots *schiltrons* (formations of pikemen) and the English knights.

Edward I and later Edward III and the Black Prince were to demonstrate the art of combining archers and men at arms. Although such tactics were only brought to a point of perfection in the early part of the Hundred Years War, the Scottish leaders understood the dangers posed by the English bowmen. It is difficult to understand, therefore, why Edward II decided to act in defiance of precedent and good sense by limiting the role of his archers to that of passive watchers. Probably he was not man enough to control his impetuous knights and allowed their arrogance to cost him the battle in the same way that the French commanders were to do 30 years later at Crécy.

The battle of Bannockburn was fought on terrain unsuited to heavy cavalry. Edward's decision to accept battle on the low-lying bogs and ponds of an area next to a tidal estuary was absurd. It went against every convention of medieval warfare. The problem may have been that he had promised to relieve Stirling by a certain date and thought so little of his enemy that he felt he could take liberties with him. But by accepting combat at Bannockburn, Edward was allowing Bruce to dictate the terms of the encounter. He was letting his opponent choose conditions which would suit his own army rather than Edward's. Even the governor of Stirling Castle, Sir Philip Moubray, who had been given a safe conduct to visit the English camp, tried to persuade Edward to wait and see what the Scots did, rather than attack Bruce in the swamps. This was wise advice, but Edward would not listen. His pride was at stake and he would not turn his back once he had sighted the Scots.

The atmosphere in the Scottish camp was calm. Bruce had divided his army into four 'battles', the first three commanded by his brother, Edward Bruce, the Earl of Moray and Sir James Douglas. The rear 'battle' he commanded himself. Each 'battle' consisted of 1,000-1,500 pikemen, who fought together almost in the fashion of a Greek phalanx. Such a formation would be difficult for the English knights to break on their own, but they could be broken by a combination of archers and knights, with the horsemen making quick darts into the spaces created by the bowmen. So apprehensive was Bruce of the dangers of the English archers that he had ordered his 500 light horsemen under Sir Robert Keith to attack them as soon as they appeared. He need not have worried, however, for archers were the last things on the minds of the English knights. They were vying with each other for honour, glory and rank.

Edward II was not just a military bungler, he was also a poor judge of men. In selecting a commander to lead his vanguard he had made just about the most ridiculous choice possible. In spite of the fact that Humphrey de Bohun, Earl of Hereford, was Constable of England by hereditary right and a bitter rival of the Clare family, Edward decided to appoint Gilbert de Clare, Earl of Gloucester, as Constable for the duration of the battle. Humphrey was an elderly and experienced warrior and regarded Gilbert as just a young upstart who needed to be taught a lesson. By the time the English army approached Stirling these two troublesome nobles had thrown the King's army into chaos.

Although it was getting late in the day Edward's knights were clearly not under his control. Seeing the Scots drawn up in the distance, Hereford and Gloucester charged at them with all their knights, each commander trying to outdo the other. The Scots must have been taken by surprise because Bruce nearly made a fatal mistake for which he was later soundly berated by his lords. He was riding alone in front of his men when suddenly Hereford's nephew, Sir Henry de Bohun, set his lance and charged straight at him. Here was a chance to make a name for himself, thought de Bohun, and to settle the campaign with a single blow. It was a famous encounter, but it did not last long. Bruce ducked aside from de Bohun's lance and hit the young knight such a mighty blow with his battleaxe that he split his helmet and his head in half. The elated Scots took this as a good omen and cheered their king as he returned from the combat, holding just the axe handle and lamenting the loss of his favourite weapon.

While Hereford's men were attacking Bruce so unsuccessfully, another English cavalry division led by Clifford and Beaumont had set off northwards to circle round the Scottish army and reach Stirling Castle. To do so they needed to pass St. Ninian's kirk. It was here that Moray's 'battle' had been placed to prevent the English trying to outflank the Scots. Seeing the Scottish pikemen advancing, Sir Thomas Grey called on Beaumont to halt. As in so many medieval disasters, when a knight's courage was called into question common sense was tossed to the winds. Beaumont told Grey that he could flee if he wanted, whereupon Sir Thomas and the rest of his knights charged wildly into the mass of pikemen and were quickly speared. In fact, although neither side knew it at the time, the decisive struggle of the entire battle was now taking place. Moray's pikemen were on their own, without the support of their fellows and without the aid of swamp or forest to guard their flanks. Yet the English knights could not break their formation without the support of archers. Eventually, the English knights were driven off and Moray's men were able to rest. But the earl had proved that the English knights alone could not hope to prevail. Unless there was a marked change in English tactics the next day the Scots would certainly win a great victory.

Morale in the English camp was by now very low. In a mood of despondency, Edward now made a further blunder, ordering his knights to cross the Bannock Burn and establish their camp there. So swampy was the ground that doors and tables had to be taken from a nearby village to take the weight of the horses. The infantry, disregarded almost throughout the campaign, consoled themselves by getting drunk. Few of the English were able to sleep that night, expecting at any moment to be attacked by the Scots. Yet curiously the confidence of the Scots was by no means as great as the English may have thought. Having given the English king a 'bloody nose', Bruce was thinking of withdrawing his troops to continue a guerrilla war rather than risk everything on one day's fight. Apparently, it was the news that an English knight – Sir Alexander Seton – had deserted that convinced him English morale was broken and that he should fight the next day.

As day dawned the English knights now tried to deploy themselves to face a possible Scottish attack. But they could not find a firm footing anywhere and the improvised walkways of doors and tables could only bear the weight of one horse at a time. Crossing the burn by a number of fords the knights reached some firm ground at last, but on a very narrow front. Here the Scots attacked them. It might be thought that no one in their right mind would use heavy cavalry on so constricted a front, where the horses could not even accelerate to a trot, but this was where the English eventually consented to fight. With their horses constantly vulnerable, the English knights had no answer to the massed pikes of the Scots.

It was now that the archers were belatedly brought round to begin firing on the Scottish masses. With no possibility of escape the Scots pikemen began to fall in large numbers. Had Edward used his bowmen like this from the start and protected them with his knights the outcome of the battle would have been very different. But now, firing from an exposed position and with no protection, the English archers were swept off the field by Sir Robert Keith's light horsemen. This was virtually the end of Edward's hopes, and Sir Giles d'Argentan with 500 knights helped to lead the king away from the field and to safety. Having done this, and determined to justify his third place ranking, Sir Giles rode back into the battle and died on the Scottish pikes.

On seeing the king flee, the English army broke up completely. But as many of the knights were pushed back they began to realize to their horror that there was no real escape. In front were the Scottish pikes, to the rear a treacherous ditch of the Bannock Burn, and on their flanks either swampy ground or the estuary of the River Forth. Panic became widespread and knights, archers and infantry fled in all directions, many drowning in the deep water of the burn. The triumphant Scots pillaged the English camp, in which Edward lost his shield, his seal and all his personal effects. Not since the battle of Hastings had an English king been so comprehensively defeated. And yet as his son Edward III was to demonstrate against the Scots at the battle of Halidon Hill in 1333, the right tools in the right hands guarantee the right result. Edward II may not have been the effeminate weakling of Marlowe's portrayal, but as a military commander he was a disaster.

THE BATTLE OF THE YELLOW FORD (1598)

In Elizabethan times the English troops in Ireland were often isolated in small garrisons, cut off from the support of their fellows in Dublin Castle. To convey supplies and reinforcements to these men it often needed almost the whole army as escort. This played into the hands of able Irish leaders, notably Hugh O'Neill, the Earl of Tyrone, who would try to draw the English into terrain suited to the Irish style of warfare. Siting small bases in the wilderness thus placed the English at a dangerous disadvantage since it rendered them liable to be attacked while on the march. One such fort was that built on the River Blackwater by Thomas Lord Burgh in 1597. Burgh saw the fort as 'an eyesore in the heart of Tyrone's country', and as a key to an English victory over the rebel Earl. But Burgh was wrong, perhaps confused in his mind and certainly dying of 'Irish ague'. In fact, the positioning of the fort was a dreadful strategic error which would lead to one of England's worst defeats at the hands of the Irish.

Burgh had planned the fort to be a symbol of England's strength, but it was a paltry thing, of a rough earthwork construction, garrisoned by 150 men under Captain Thomas Williams. Conditions in the fort were primitive and Williams often had to fight O'Neill's men to get fresh water from the river nearby or wood for his fires. The fort was clearly a military liability. Yet to evacuate it would be to lose face and this the English were not prepared to do. Nor was O'Neill prepared to capture it, for to do so would be to lose his bait to draw the English army on to disaster.

During the winter of 1597–8 Williams had great difficulty in keeping the fort standing, as a section of the walls was washed away by flooding. He was also running short of supplies. An expedition would have to be launched to reprovision the fort and this was to be led by Sir Henry Bagenal, O'Neill's brother-in-law and bitter enemy. It would probably have been best to abandon the fort, but in the eyes of the Queen and her Council it had become a symbol which could not be surrendered or lost.

Bagenal set out from Armagh for Blackwater Fort with 4,000 infantry and 300 cavalry, of whom the 1,500 recruits recently arrived from England were particularly poor material, selling their arms and uniforms as they received them and deserting as soon as the opportunity arose. Bagenal was experienced in the ways of Irish warfare and knew that he would have to fight to reach the fort. The main point was to avoid the 'great plashes' and the boggy terrain where the Irish would be deadly. His column stretched a mile and his infantry was divided into six regiments of 500 men, each marching 100 yards apart. Meanwhile O'Neill, who was waiting for him with 5,000 men, kept up a galling fire on the English column from the flanks. The English recruits, unused to gunfire, began to panic and the gaps between the regiments grew longer and longer. Soon they were separated from each other by hills and clumps of trees.

The English army had split into three sections, each heavily engaged by the Irish. English soldiers began to panic, stumbling across cornfields and into peat bogs. Everything was going as O'Neill had planned. The battle was being fought on the very ground he had prepared, by digging pits to combat the English cavalry. The main obstacle the English faced in their effort to reach the ford leading to Blackwater Fort was a ditch of great length, some say the best part of a mile long, which was 'full of water and thorns' and flanked on each side by bogs. It would be impossible for the English to bypass this obstacle while still under fire from the Irish.

As Bagenal tried to rally his men he raised his visor for an instant and was shot in the face. With the death of the commander, it was left to Thomas Wingfield to organize a retreat towards Armagh. In the confused fighting the English lost 25 officers and 800 men killed, as well as a further 400 wounded. In addition, some 300 of Bagenal's soldiers deserted to the rebels. In England the news of the battle was received with horror; it was the worst defeat ever suffered at the hands of the Irish. The Queen wanted to know how it could have happened. Her commander in Ireland, the Earl of Ormond, blamed Bagenal for marching with his regiments too far apart. Others blamed the cowardice of the recruits, many of whom had fled at the first shot, throwing away their arms and armour. But everyone missed the point. It was the defence of Blackwater Fort that caused the disaster. It was not worth risking so much for a symbol which had no strategic value at all. For the English the only positive outcome of this disastrous episode was that Captain Williams was forced to evacuate the fort through lack of supplies. The responsibility should not have been his. The mistake was made by those who built the fort in the first place.

CHAPTER 5: THE BUTCHERS

Lapses of the Great

Butchery – the avoidable human cost of war – is often associated with the trenches of World War I. Yet even great commanders such as Napoleon Bonaparte and Frederick the Great were responsible for wasting the lives of their men on a massive scale. Napoleon once said, 'A man such as I am is not much concerned over the lives of a million men.' In fact, the wars fought by the Corsican were to cause the deaths of at least that many. His boundless ambition had a high cost in human lives. Nor was the ambition and the ceaseless campaigning of Frederick the Great of Prussia proportionately cheaper. A number of his generals appear to have shared 'Old Fritz's' lack of concern for casualties.

Cannon fodder

During the Seven Years War, Frederick made a terrible mistake at the battle of Leignitz in 1760, a battle that he eventually won – though at an exorbitant cost. Outnumbered 53,000 to 44,000, Frederick had split his army into two parts, commanding one himself and leaving the other to Field Marshal Zieten. Hearing the sound of cannon carried by the wind, he thought that Zieten was already engaged, without his help. In fact, it was just a brush between skirmishers. Panicking, Frederick shouted, 'My God, Zieten's attacking already and we're still a league and a half short! What's going to happen? My infantry's still not there!' He stopped his march and sent forward ten battalions of grenadiers out of their wooded cover and straight into the centre of the Austrian army. With no support whatsoever they were mown down, 5,000 falling in the space of 30 minutes: here was butchery on the scale of the Western Front in 1914-18. In spite of a victory won for him by Zieten's determination, Frederick's losses of 16,670 out of a force of 44,000 formed a higher proportion of losses than in any other of his battles.

But at Kolin in 1757 Frederick's mistakes led to a crushing defeat. So senseless was the slaughter that a modern historian, Christopher Duffy, has described the battle as 'for the Prussian army, what the Charge of the Light Brigade represented for the British in the nineteenth century'. Frederick, with 35,000 men, was again outnumbered by Marshal Daun's 53,000 Austrians, who were situated on Krzeczhorz Hill, protected by batteries of heavy cannon. In spite of the best efforts of Prince Moritz of Anhalt-Dessau to persuade Frederick to change an apparently suicidal order to his nine battalions to make a frontal assault on the hill, 'Old Fritz' was insistent. As the Prussians marched up the slopes the Austrian cannon cut swathes through them, inflicting 65 per cent casualties. At Kolin, Frederick's losses amounted to 10,000 killed or wounded and 5,000 prisoners.

It is instructive to study the cost of Frederick's campaigns during the years 1756 to 1762. Prussia, in population the smallest state of the participants in the Seven Years War, suffered over 180,000 war deaths, and had 62,000 men taken prisoner. One regiment lost nearly 4,500 men, which meant that it had effectively been wiped out three times

Frederick the Great was one of the undisputed great captains of history. But his readiness to countenance losses even greater than his enemies in pursuit of victory earned him the nickname 'Grave Digger' from his men.

over. The entire population of Prussia suffered a reduction of half a million in population during these years. Such was the profligacy in human lives of Frederick the Great.

Napoleon's ossuary

We have already seen that as Napoleon got older his health declined to the point where it undoubtedly affected his military judgement. After 1807 there are far more occasions when he was prepared to use brute force to achieve a breakthrough. A prime example of this was at the hard-fought battle of Wagram in 1809. Failing to manoeuvre a decisive advantage on the first day of the battle, he adopted a 'battering-ram' technique using Marshal Etienne Macdonald's corps of 8,000 men, who were formed into a gigantic column, in the shape of a kind of hollow rectangle, and then ordered forward to break the Austrian line. For the massed Austrian cannons Macdonald's men formed an unmissable target. Napoleon was trying to 'buy' a breakthrough with the lives of his men, rather than earn it through his tactical skills. Macdonald's losses were enormous – over eighty per cent casualties – and the breakthrough failed. Napoleon had lost control of the situation to such an extent that he actually ordered General Wrede to try to bail out the desperate Macdonald with the unheard-of order, 'do as seems to you best'. As it happened, the martyrdom of Macdonald's men had been unnecessary. On both right and left flanks

Marshals Davout and Masséna had broken the Austrian line. Nevertheless, it had been a close thing.

At Heilsberg in 1807 everyone on the French side seemed to be off form at the same time. Napoleon was determined to defeat the Russians under General Bennigsen once and for all and had decided on his course of action regardless of his enemy's position – a fatal mistake. The tempestuous Marshal Murat, with the cavalry reserve, was swept away by the Russian horsemen and when General Savary came to his aid and saved the situation, he was rewarded with the rough end of the Marshal's tongue and an accusation of cowardice. Savary responded that he wished Murat had less courage and more common sense. Marshal Soult's tactics of rushing his divisions head on at the Russian defences and battering away without any attempt to manoeuvre was meat and drink to the resolute but much less skilful Russians. Napoleon was getting nowhere with his attacks against a numerically stronger army in a powerful position. However, instead of calling off the attack he kept feeding in more men to the attritional struggle in the centre. The arrival of Lannes and his fresh divisions should have improved matters, but Napoleon allowed the headstrong Marshal to squander his troops in a disastrous night attack, which cost him over 2,000 casualties to no purpose. In all, Napoleon lost over 10,000 men at Heilsberg in one of the poorest tactical displays of his whole career. Bereft of ideas, he fell back on attrition against an enemy who thrived on that kind of contest.

It is hardly surprising that commanders occasionally lose their tempers under the constant stress of campaigning. But the results of such lapses can be disastrous, as at the minor battle of Somosierra, fought by Napoleon in Spain in 1808. En route to Madrid, Napoleon found his path blocked by the troops of General San Juan. Many of the Spaniards were no more than peasant levies and were no match for Napoleon's regular troopers. But San Juan had taken up an immensely powerful position at the head of a pass, through which wound the road that Napoleon had to take to reach the Spanish capital. The Emperor was so scornful of his enemy that he seemed to want to demonstrate his contempt for them. Although French infantry had been sent into the hills and were working their way towards the Spanish position, Napoleon lacked the patience to wait for them to engage. In a fit of pique he suddenly turned to the commander of his personal escort, the colonel of the Polish Light Horse, and ordered him to charge the Spanish position. Here was an early version of the Charge of the Light Brigade – a mere 87 troopers with no other support ordered to ride up a road towards a mass of Spanish guns. After a few salvos had shot down half the force, the survivors took shelter. But Napoleon could not allow this and sent a messenger forward saying, 'My Guard must not be stopped by peasants, by mere armed brigands.' Thus the wretched Poles had to continue their charge, suffering 60 casualties out of the 88 starters and failing to reach the Spanish lines. Whether through shame or otherwise, Napoleon failed to report these losses in his communiqué of the battle, which was eventually won by more conventional tactics.

Pickett's charge

Although the reputations of Civil War generals like Ulysses S. Grant and Robert E. Lee are proof against accusations of incompetence, both were guilty of squandering the lives of their men on occasions. At Gettysburg in 1863, Lee ordered General Longstreet to

'Pickett's Charge' at the Battle of Gettysburg in 1863 was the climax of a massive frontal assault on the Union defences across open ground ordered by Confederate commander Robert E. Lee. Its bloody repulse is seen by some as the turning point of the American Civil War.

launch a massive frontal assault on the Union position at Cemetery Ridge. Even though his subordinate warned him that it was an impossible target, Lee insisted because he 'thought his men were invincible'. What followed was a heroic but forlorn charge of 15,000 Confederate veterans led by General Pickett across more than a thousand yards of open fields against massed artillery and entrenched infantry on Cemetery Ridge. Longstreet was distraught: 'I could see the desperate and hopeless nature of the charge and the hopeless slaughter it would cause.' Pickett's charge has become an American epic, yet it was also a turning point in the war and a tragic waste of the Confederacy's best troops. Half of the 15,000 did not return and Pickett's own division suffered nearly seventy per cent casualties, including three brigadiers and all its thirteen colonels. Lee was man enough to take the blame himself. He had been suffering from diarrhoea and his judgement may have been impaired. The same can certainly be said for Union General Grant and his fit of pique at Cold Harbor the following year.

Like Lee at Gettysburg, Grant made the mistake of thinking his men stronger in spirit and morale than his opponents. On 3 June, 1864, frustrated by the seemingly endless attrition of the trench fighting outside Cold Harbor, his temper snapped. In a fierce exchange of argument with his fellow commanders he promised that he would break through the Rebels' defences through sheer brute force, and on that very day. At once he ordered a frontal attack on the Confederate lines. But Grant's men knew that they were going into a situation as dangerous as the Marye's Heights at Fredericksburg (see p. 116).

During a day of head-on assaults, the Union lost 7,000 men in casualties, most in the first ten minutes. It was almost a pre-echo in miniature of 1 July, 1916 on the Somme (see p. 114). One of Grant's colonels wrote later, 'I am disgusted with the generalship displayed. Our men have, in many instances, been foolishly and wantonly sacrificed.'

Bloody Incompetents

So far we have been concerned with the occasional profligacy, immediately regretted, of great commanders. However, there have been butchers aplenty who have no right to be considered even competent generals and whose contempt for their own troops allowed them to squander their lives without a thought of self-recrimination.

The siege of Plevna

During the siege of Plevna in the Russo-Turkish War of 1878 the Turkish town was defended by General Osman Pasha with 22,000 men and 58 guns. On 30 July 1878 the Russian generals Krudener and Shakofsky launched a massive assault with 35,000 infantry, supported by 170 guns. They ordered a purely frontal assault against skilfully constructed defences, without even reconnoitring the Turkish position. Battalion after battalion walked heads down into a maelstrom of shot and shell. The Russians were completely routed, losing over 7,300 men in a single day. A British observer wrote:

> They seemed to have no advance-line of skirmishers. Serried ranks of infantry – three battalions, I believe – climbed in a solid body the bank of the last ditch and advanced in a line parallel to the redoubt. The attack was purely frontal. Hardly had they appeared when a dozen bugles sounded 'Fire' and a terrific fire, coming from three sides, brought the enemy to a dead stop. The survivors surged back and were swallowed up by a second, which had meanwhile commenced to advance. A third followed at a short distance . . . The Russians surged forward and recoiled from the slope like the waves of a tempestuous ocean.

In three weeks of such attacks, the Russians lost 27,500 men against an enemy with a reputation for poor marksmanship.

Kamikaze

During the Russo-Japanese War of 1904–5, the Japanese commander, General Nogi, used 'human wave' attacks during the siege of Port Arthur on the Chinese coast. The patriotic fervour of the Japanese soldiers made them ready and eager to make sacrifices which helped to conceal the incompetence of their commander. Not all of Nogi's fellow generals agreed with his plan for a frontal assault on so powerful a fortification. Yet when they suggested postponing the attack Nogi angrily overruled them. Nogi had been told that 10,000 lives would be a cheap price to pay for success, but before the city eventually fell, six times that many Japanese would fall.

As soon as Nogi's grand assault was launched on 19 August 1904, the faults in his preparations were revealed. The houses with glass windows on the exterior walls of the city – identified by the feeble Japanese reconnaissance work – in fact turned out to be heavy fortifications of steel and concrete. Before the attack was repelled the Japanese had suffered 16,000 casualties because of Nogi's insane idea that Port Arthur could be easily taken by infantry assault.

'You will be able to go over the top with a walking stick, you will not need rifles. When you get to Thiepval you will find the Germans all dead, not even a rat will have survived.'

Brigadier-General addressing Newcastle Commercials 1 July 1916 on the Somme

The Germans were not dead. They had been in deep concrete bunkers throughout the British bombardment and emerged once the firing ceased to man their machine guns and shoot down the advancing British infantry. Some British officers advanced with walking sticks, others with umbrellas, some even kicked footballs. They were shot down just the same. There are no records of casualties to rats, though they were thought to be light.

Hunter-Bunter

The Gallipoli campaign of 1915 saw many examples of needless sacrifice of troops, ordered by incompetent British generals. The most notorious of these was Major-General A. G. Hunter-Weston, dubbed the 'Butcher of Hellas' by historians and 'Hunter-Bunter' by the troops, because of his ample girth which reminded them of the comic fat boy Billy Bunter. Apparently a charming man away from the battlefield, Hunter-Weston was a heartless butcher anywhere near one. As the commander of Britain's élite 29th Division he remarked, 'Casualties? What do I care for casualties?' On another occasion he told an officer whose brigade had just lost over 1,300 casualties, that he was pleased to have 'blooded the pups'. Before landing at Cape Hellas in April 1915, he had encouraged his troops by telling them that they must 'expect heavy losses by bullets, by shells, by mines and by drowning'. Having got his men ashore, Hunter-Weston now told them that 'every man must die at his post rather than retire', even though the weakened Turks were retreating from the massive British landings. In the time before he collapsed under the pressure of command, Hunter-Weston launched a series of attacks on the hills of Achi Baba and Krithia which exhausted three whole British divisions. He rejected the idea of using surprise, or setting off at dawn or dusk, and seemed obsessed with attacking frontally and in broad daylight, thereby ensuring maximum casualties.

Russell's Top

The film *Gallipoli* has immortalized the Australian assault at Russell's Top on 7 August 1915, during the campaign against Turkey in World War I. Six hundred men of the

Australian Light Horse Brigade were sent over in four waves to attack a Turkish-held hill known as Baby 700, near Lone Pine. The hill was protected by trenches and flanking machine guns. Inexplicably, the naval barrage intended to keep the Turks under cover prior to the assault had stopped seven minutes early. The Australians were uncertain as to whether they should attack immediately or wait. They feared that if they attacked the naval guns might open up again. Instead they chose to wait for the seven minutes to elapse, giving the Turkish defenders time to get ready for them.

As the first wave of 150 men went over the top they were hit by a hail of bullets from a range of just 60 yards. Few men even got half-way and the whole line was wiped out. Within two minutes the second wave was ready to go, even though they could see that the first wave had failed completely. As they went over the top they too were struck down within seconds. The third wave, under Major Todd, now prepared to follow. Todd pointed out the wasteful futility of the assault and his comments were passed to Colonel Antill. But Antill had heard a rumour than an Australian flag was flying in the Turkish trenches and now ordered Todd to provide support for the plucky but fictitious Australians who had reached the Turkish line. The third wave duly went over the top and were wiped out. The fourth wave also complained that the task was impossible but again they were sent to their deaths by Antill, without an explanation why. In the four attacks 435 men were lost, with 232 killed, and all because no officer had the courage to stop the pointless slaughter.

'The men are in splendid spirits. Several have said that they have never before been so instructed and informed of the nature of the operation before them. The wire has never been so well cut, nor the artillery preparations so thorough.'

General Douglas Haig, 1 July 1916

In many places the wire was not cut. The artillery had failed. Thousands of lives would be lost because the men could not break through the barbed wire and had no cutters with them. In other places the Germans concentrated all their firepower on where the wire was cut, knowing the British must come that way.

'Never So Well Cut . . .'

It is fitting to end our study of military profligacy by looking at the blackest day in the history of the British Army and possibly of any army – 1 July 1916, the first day of the battle of the Somme. By nightfall the British had suffered 57,470 casualties out of the 120,000 men who had left the trenches that morning, with some 21,000 men killed, most in the first 30 minutes of the attack. No fewer than twelve divisions suffered over 3,000 casualties each and the 1st Hampshires were virtually wiped out, with no one left at the end of the day to describe what had happened. The 10th West Yorks were annihilated in less than a minute. Many of Hunter-Weston's 29th Division were caught in the open, where the barbed wire was not cut, and machine-gunned at will.

The men chosen to carry out the assault had been promised that the barbed wire had been cut by the enormous artillery bombardment and that the only Germans they would see would be dead ones. While some officers armed only with swagger sticks or umbrellas led their men forward, others, like Captain Wilfred Nevill of the East Surrey Regiment, kicked footballs towards the German lines and charged after them. The Germans, hidden in deep concrete bunkers throughout the bombardment, had ample time to reach their trenches before the British troops could cross no man's land. Despite the heavy bombardment, the wire had remained largely intact and in the few places where it had been cut the Germans had made certain to cover the gaps with machine guns. When Hunter-Weston, commanding VIII Corps, told his men the wire had been blown away, many of them could see that it was still standing and that their commander was lying through his teeth.

The decision by Hunter-Weston to order the heavy artillery to end their barrage ten minutes before the troops went over the top, and the field artillery two minutes before, was a direct cause of many of the British casualties that day. The artillery should have continued firing until after the British soldiers had left their trenches. The result was that the Germans had ample notice that the assault was about to take place and plenty of time to emerge from their concrete bunkers to set up their machine guns. This single blunder by Hunter-Weston cost VIII Corps 14,000 casualties. Even so, Haig was so ill-informed that he claimed that 'few of the VIII Corps left their trenches'. Conscious of his own mistake, Hunter-Weston hastened to inform the Chief of the Imperial General Staff, General Robertson, that the artillery had let him down badly, failing to cut the wire and suspending their barrage too early.

Haig's 'Big Push' – as he had entitled the Somme offensive – eventually got bogged down into the kind of attritional battle that was typical of the period 1915–17 on the Western Front. But the single day – 1 July 1916 – will remain for the British Army the 'black day' – rather at odds with Haig's callous summing up: 'The general situation was favourable.'

'They advanced in line after line, dressed as if on parade, and not a man shirked going through the extremely heavy barrage, or facing the machine-gun fire that finally wiped them out . . . I have never seen . . . such a magnificent display of gallantry, discipline and determination. The reports I have had from the very few survivors of this marvellous advance bear out what I saw . . . that hardly a man of ours got to the German front line.'

Brigadier-General Rees, GOC 94th Infantry Brigade of 31 Division, 1 July 1916

The 'marvellous advance' was a complete disaster; the 'magnificent gallantry' wasted. The men had no chance whatsoever. Casualties were nearly total.

Ambrose Burnside's decision to attack the massive Confederate defences at Fredericksburg on 13 December 1862 was probably the greatest blunder of the American Civil War. Here the Confederate defenders are protected by a wall in front of which it is said that the bodies of Union soldiers were piled high.

THE BATTLE OF FREDERICKSBURG
(1862)

Laurel and Hardy never made a horror film, but if they had, and fat, smiling Ollie had played Ambrose Burnside presiding over an abattoir scene, it would have been set on Marye's Heights, the ridge behind Fredericksburg, on 13 December 1862. Here Burnside committed the most atrocious blunder of the American Civil War – and no one laughed.

When President Lincoln finally lost patience with the slow-moving McClellan as commander of his army and selected Ambrose Burnside instead, he was making a big mistake and everyone, including Burnside, knew it. The new commander tried hard to stop Lincoln, saying, 'I do not want the command. I am not competent to command such a large army.' But Lincoln was insistent and Burnside eventually gave way, apparently celebrating in

champagne with his fellow commanders. The President wanted action and Burnside was prepared to provide it, whatever the cost.

His army, now 113,000 strong, outnumbered Lee's 80,000 and Burnside was determined to make this count. He ordered his fellow commanders to prepare for a crossing of the Rappahannock River opposite the town of Fredericksburg. He could hardly have made a worse choice. Here the river is wide, deep and fast-flowing in winter. Beyond the town of Fredericksburg there is a high ridge, out of artillery range for the Union guns, but perfect as a defensive position overlooking any crossing point of the river. Confederate general Robert E. Lee occupied these heights in massive strength, hardly believing that Burnside could seriously think of crossing the river here. Burnside was contemplating just that – but had to sit waiting for five days because of a mix-up over the ordering of pontoon bridges. It would have been better for everyone if they had never arrived.

When the pontoons did arrive Burnside was ready to cross, but, on hearing that the Confederates had occupied the heights opposite the river, he was visited by a fit of indecision. The responsibility was too much for him and he spent sleepless nights, tossing about trying to steel himself to give the order to cross. On the night of 12 December he wandered listlessly through his camp, like some latter-day Henry V before Agincourt. Eventually, from somewhere, he dredged up the suicidal idea of a full frontal attack on forces commanded by the most able general of the age, across a wintry river full of drifting ice, against massive fortifications on a forbiddingly steep ridge.

When the Union soldiers learned of the plan their response was poignant but realistic. Thousands of them wrote their names and addresses and next of kin on their handkerchiefs and sewed them onto the back of their tunics before they advanced, so that their bodies could be identified later. Others wrote farewell letters to their parents and wives, in which they revealed that they knew their task was hopeless and that they would be fortunate to survive such an attack.

After the pontoon bridges had been positioned, the Union forces crossed the river in two massive divisions, Franklin's 50,000 men attacking downriver of the town of Fredericksburg against 'Stonewall' Jackson, while Sumner's and Hooker's men faced the unenviable task of assaulting Longstreet's troops on Marye's Heights. A Confederate artillery officer summed it up when he said, looking over the open fields the Yankees would have to cover before reaching the heights, 'A chicken could not live on that field when we open up.' With over three hundred guns primed and ready, he was not exaggerating.

On Marye's Heights, Longstreet's men held one of the strongest defensive positions imaginable, tipped with a stone wall half a mile long behind which were four ranks of Georgians and North Carolinians, who kept up so fast a fire that it has been likened to the effect of machine guns. Time and again the Union troops charged up towards the wall only to fall, riddled with bullets. In this small stretch of battlefield five miles in width 6,000 Union troops fell. Watching the proceedings through glasses, Burnside had no answer but to order more attacks. Fourteen times the assault on Marye's Heights was renewed until Hooker, on his own authority, called his men back. Bitterly he turned on Burnside, 'Finding that I had lost as many men as my orders required me to lose, I suspended the attack.'

By this stage Ambrose Burnside was no longer smiling. He was raging at his generals as 'cowardly scalawags' and his men, 15,000 of whom now lay dead or wounded in the fields across the river, as 'thieves' and 'poltroons'. He came to another decision. He would ride up the Marye's Heights at the head of his old IX Corps. Deterred by his generals, he now collapsed in tears, moaning 'those men, those poor men'.

On the other side of the river the Confederates were beside themselves with joy. 'Stonewall' Jackson came up with a plan so bizarre that even Burnside would have thought twice about it. He suggested a night attack, in which Lee's entire army would strip naked – to help distinguish them from the Union troops in the darkness – before swimming the river and rushing at Burnside's shattered army. Lee's prudishness, or common sense, prevented such mad antics in the middle of a snowy December night.

The consequences of the battle of Fredericksburg were immediate. Mutiny broke out among the Union Corps and Divisional commanders when Burnside proposed another river crossing and frontal assault. Hooker warned that he could not guarantee that his troops would obey such an order, whereupon Burnside promptly sacked him. Hooker sent one of his officers to complain to Lincoln. Burnside then went to see Lincoln himself, accusing Hooker of cowardice. Lincoln, weighing the evidence of Fredericksburg and the likely cost of keeping Burnside in a job, came down on the side of Hooker. Burnside was sacked and Hooker got his job. With whom does the blame lie for Fredericksburg? Burnside never tried to shirk his responsibility, but he had after all warned the President that he was not up to the job. In the final analysis the buck stopped with Lincoln.

THE BATTLE OF GRAVELOTTE-ST. PRIVAT (1870)

General Friedrich von Steinmetz, a disciple of Marshal Blücher, was as unorthodox an officer as any in the Prussian army of 1870. A confirmed rebel even at 74, he wore a non-regulation black oilcloth cap over his white hair. He had enjoyed bloody successes in previous wars, notably against the Austrians in 1866, and was never one to trouble himself greatly over casualties. As a previous commander had noted, he needed 'to be kept under control'. As he grew older his behaviour became more violent and unpredictable, notably after the death of his beloved only daughter in 1854. Apparently for the rest of his life he suffered visions in which the girl appeared to him. These reminders of what he had lost gave him a wild and driven personality, often at odds with his military superiors and contemptuous of civilian interference. So wilful had he become by 1870 that he found it unbearable to follow the orders of the Prussian commander-in-chief, General Helmuth von Moltke. As a result of his insubordination his troops suffered heavy casualties at Spicheren, and at Gravelotte he almost snatched defeat from the jaws of victory in the decisive battle of the Franco-Prussian War.

Steinmetz commanded the Prussian First Army in 1870 and, on 6 August, struck directly at a French corps occupying a ridge at Spicheren. Ordered to make frontal assaults against the powerful French defences, Steinmetz's men were shot down in thousands. Only the fact that every Prussian commander in the vicinity came to his aid saved the Prussians from a heavy and unnecessary defeat. Nevertheless, it was a Pyrrhic victory, with Prussian losses twice as high as those of the French. Moltke was furious with Steinmetz, but the old man was not prepared to listen to any criticism. As he moved his troops towards

Gravelotte he was determined to conduct the war in his own way. What had been good enough to clear the Corsican Bonaparte from Germany in 1813 would do to beat this mountebank emperor, Napoleon III, in 1870.

The battle of Gravelotte-St. Privat on 18 August 1870, was to be a real test of strength between the French and Prussian armies. Both sides were fully committed to the encounter and over 300,000 men and 1,200 guns would be deployed on a battlefield seven miles wide. Many commanders failed to meet their responsibilities on this day, but none more so than Steinmetz. His orders from Moltke were for his First Army to fight a holding action against Frossard's II Corps and Leboeuf's III Corps on the extreme right, or southern end of the French line, near Rozerieuilles. The crucial role in the battle was given to Prince Friedrich Karl's Second Army, whose task was to turn the left flank of the French army between Amanvillers and St. Privat. In the event, however, just about everything went wrong for the Prussians.

Depressed by his 'holding' role, Steinmetz decided to disobey his orders and make a series of mass attacks on the steep ridge facing him. Although Moltke had specifically withdrawn Goeben's VIII Corps from Steinmetz's control, this did not prevent the old man from ordering Goeben to attack the ridge down a narrow passage between heavy woods and steep slopes. The French concentrated all their fire on this narrow front and blew Goeben's troops to pieces. Not

The Franco–Prussian War of 1870–1 revealed how much modern artillery and rifle fire had improved. Attacks across open ground, as here by the Prussian Guard at St. Privat, were bloody and expensive failures.

content with wrecking just once corps, Steinmetz now ordered VII Corps, complete with their artillery, and even the First Cavalry Division to try to force their way along the narrow road and through the shattered remnants of Goeben's men. The result was little short of a massacre. When the cavalry recoiled from the massed French artillery and rifle fire it was a signal for the collapse of Steinmetz's entire front. Here was an opportunity for the French to sweep down from the ridge and crush the German First Army. Leboeuf pleaded with Marshal Bazaine to order such an attack, but Bazaine could not cope with the enormity of the pressure he was under. Incredibly, at the time when an order from him might have turned the whole war in France's favour he concerned himself with the siting of an artillery battery on Mount St. Quentin, guarding his left flank. With the Prussians streaming away in defeat Bazaine had 'abandoned his post in face of the enemy'.

On the Prussian left, at St. Privat, Prince Friedrich Karl was showing the prudence that had been so lacking in Steinmetz. However, all his careful garnering of troops was about to be ruined by the extraordinary action of Prince Augustus of Würtemburg, commander of the élite Prussian Guard Corps of 30,000 infantry. Inexplicably, and entirely on his own responsibility, the prince ordered the guards to make a frontal assault on the French positions at St. Privat, without the benefit of any artillery support. Even when General von Pape pointed out that the French positions were quite undamaged, he was told to obey orders and begin the attack. Led by their mounted officers, the guards marched shoulder-to-shoulder and in impeccable order up the slope against the French guns. So heavy was the fire from the French chassepot rifles that survivors spoke of walking head down as if through a blizzard or hailstorm. No one got closer than 600 yards from the French positions and the Guard Corps suffered 8,000 casualties in just twenty minutes. Even Steinmetz was never responsible for butchery on this scale.

Only the limpest possible leadership by Marshal Bazaine prevented the French winning this great battle. Their subsequent retreat brought little feeling of victory to the Prussians, who had seen the cream of their army wasted by incompetent blunderers. Prussian casualties of 20,163 had to be set against French losses of just over 12,000. King Wilhelm I of Prussia was disconsolate at the squan-

dering of his Guards, while Bismarck raged against the butchery of General Steinmetz. Rarely has a victorious army made mistakes to match those of General Steinmetz and Prince Augustus of Würtemburg. Steinmetz's career was over. He was relieved of his command and sent into retirement as Governor of Posen – a location felt to be far enough away from the centre of operations for safety.

THE BATTLE OF LOOS (1915)

The British High Command did not want to fight a battle at the time or the location at which the battle of Loos took place. As it happened, the battle contained possibly the most pointless assault of World War I – and the competition for this dubious distinction could hardly be stiffer. Essentially the British commander, Field Marshal French, agreed to fight at Loos in September 1915 because the War Minister, Lord Kitchener, felt that he must support French plans to attack the German-held Noyon Salient. As a result, six British divisions were earmarked to launch an attack amidst the slagheaps and pit villages of a coal mining area, in which the Germans would have every advantage of terrain and observation. The attack was to be preceded by the release of chlorine gas, though this was not a supplement to an artillery bombardment but a substitute for the lack of heavy guns.

At 6 am on 25 September 75,000 men waited for the signal to attack. The men responsible for turning on the gas taps were waiting for their signal too. But was the wind blowing in the right direction? Might the gas not blow back on the advancing British troops? The decision lay with General Douglas Haig, who was suffering from a mild asthma attack. After a delay of nine minutes, during which there was no perceptible improvement in the conditions, he gave the order to release the gas. Clouds of green and yellow smoke billowed out and drifted sluggishly forward before forming into a dense, hanging sheet in the middle of no man's land. At least it had not blown back. However, when the British troops went over the top they soon found themselves lost in an impenetrable fog of chlorine. In such circumstances their pre-attack briefings were of little use. In some areas targets were achieved, in others there was complete failure.

By the middle of the first day Haig was convinced that he would need to feed in his reserves. He therefore called on French to release General Haking's XI Corps, which included two of Kitchener's 'New Army' divisions, fresh from England. French was unhappy at the idea of using such inexperienced men in supporting a difficult assault, but Haig felt that 'with the enthusiasm of ignorance they would tear their way through the German line'. This was callous leadership indeed and Haig compounded the felony by assuring Haking that his men would be pursuing a beaten enemy and that they would not be used unless the Germans had already received a crushing blow. Haking's divisions were in fact thrown in to support a first-day failure. The Germans were not crushed, were not retreating and were certainly not beaten.

With the shortest possible notice the two divisions, the 21st and 24th, were ordered to march to the front, only arriving after nightfall on 25 September, by which stage they had been marching for eighteen hours in pouring rain and without food. Their officers had no proper maps and no accurate knowledge of the area in which they were to operate. At one point, outside Béthune, a military policeman stopped the 72nd Brigade because their commander did not have a pass to enter the area.

Whereas the first-day assault had been preceded by an artillery bombardment - of sorts - and a release of gas, the two divisions were to advance on the 26th without any support. Even the whereabouts of their targets was not made clear to them as there were no decent maps available. In the dark and the pouring rain, crossing old trenches often by a single plank, they had landmarks such as Hill 70 pointed out to them. Perhaps it was kinder that they could not see where they were being sent. The truth was that their targets were heavily defended trenches protected by unbroken barbed wire. The attack of the two divisions was doomed from the start.

At 11 am on 26 September the attack began. In close formation, with their mounted officers at their head, they began the slow advance across no man's land. The Germans were astounded at the sight. None of them had ever seen such dense masses advancing against machine guns. The British advanced in ten lines of 1,000 men in breadth, giving the German machine gunners easier targets than they had experienced on the practice ranges at home. The riflemen actually climbed out of the trenches and stood on the parapets, triumphantly shouting as every bullet took its toll. When they reached the barbed wire, which was nineteen feet thick and four feet high, the British soldiers tried to cut through it using feeble hand clippers, quite inadequate for the task. Others ripped at the wire with their bare hands or ran up and down the line trying to find a way through. Whatever they did the result was just the same: they were shot down in their thousands. So hideous was the massacre that many Germans even stopped firing as the pitiful remnants of the two divisions retreated to their own lines. Out of 10,000 officers and men who began the attack, 385 officers and 7,861 men ended up as casualties. The Germans suffered no casualties at all. This is possibly the biggest discrepancy in casualties in all military history. The pointless folly of Loos was a grim forerunner of yet greater slaughter to come on the Somme in 1916 (see p. 114) and at Passchendaele in 1917 (see p. 89).

'It was a new device in warfare and thoroughly illustrative of the Prussian idea of playing the game.'

Sergeant Reginald Grant, on the German use of chlorine gas at the Second Battle of Ypres in 1915

Although he was not an officer, Grant's comments shows how far down the notion of British 'fair play' had permeated. To British commanders stabbing an opponent in the belly with a bayonet was far more sporting than asphyxiating him with gas. The Germans had no such scruples and were always critical of British attempts to reduce war to the level of a game.

CHAPTER 6: NAKED INTO BATTLE

Dressed to Kill

Military uniforms have served a number of purposes over the years: helping to identify one's own side from the enemy, instilling pride in one's unit or country, or even emphasizing superiority over one's adversary. The need for uniforms to conform with military tasks, or to suit environmental conditions, has until relatively recently been a very low priority. As a result men have been sent into battle wearing uniforms not only inappropriate but even capable of providing help to the enemy. The white bands across red tunics favoured by the British army in the eighteenth and nineteenth centuries gave the snipers a perfect target for hitting the heart, while the silver badges worn on their helmets by the Hessians in the American War of Independence provided a target for the brain. At Gallipoli, white armbands worn for identification at night helped the Turkish snipers pick off the British troops on the beaches. Uniforms which are warm in the winter, cool in the summer, hard-wearing and yet not too conspicuous only became the norm quite recently.

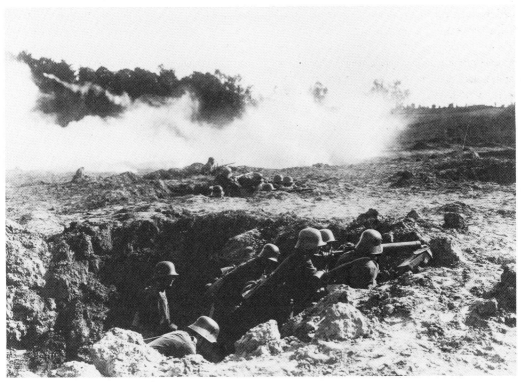

Shrapnel wounds to the head were a major cause of death during World War I, and yet the French were slow to issue steel helmets to their troops. The British helmet was markedly inferior to the German model, which covered far more of the head and neck. These German troops are manning shell holes at the Battle of Champagne in 1917.

Le Pantalon Rouge

In the decade before 1914 moves had been made by most European armies towards less conspicuous uniforms, the British setting the pattern with their khaki and the Germans adopting field-grey. Last to see the wisdom of these changes were the French, who thought of their uniforms of red kepis (caps), red trousers, blue jackets and blue great-coats as quintessentially Gallic. During the Balkan Wars of 1912–13, the French Minister for War, Adolphe Méssimy, saw the advantage of dull colours for uniforms which blend in with the background. He decided that the French armies must have such uniforms in the future. But reaction inside France was immensely hostile and he was accused of wanting to dress French soldiers in 'muddy, inglorious' colours. As he pointed out, he was simply trying to save their lives. With the increased range of rifles by 1914 inconspicuous uniforms were of paramount importance. But French newspapers pilloried him for suggesting a change, claiming that to get rid of 'all that is colourful, all that gives the soldier his vivid aspect, is to go contrary to French taste and military function'. A former minister, Alphonse Etienne, declared, 'Eliminate the red trousers? Never! Le pantalon rouge c'est la France.' Méssimy knew that he was beaten and wrote bitterly, 'That blind and imbecile attachment to the most visible of all colours will have cruel consequences.' The French even insisted on keeping their soft headwear, ridiculing the German pickelhaube (spiked helmet), which saved the lives of thousands of Germans.

'My friend, we shall not have time to make them. I shall tear up the Boches within two months.'

General Joseph Joffre, November 1914, on the suggestion from Colonel Penelon that the troops be equipped with helmets

The gaudy French uniform of 1914 cost them hundreds of thousands of unnecessary casualties. The refusal to wear a helmet, in contrast with the Germans who wore pickelhaubes (found amusing by the French), was based on sheer chauvinism – and they paid a high price for looking smart.

During the battle of the Marne in 1914 the folly of the French was clearly revealed – in more senses than one. The 246th Regiment had to cross an exposed plain on a bright, sunny day. Against the light yellow background of the fields the French uniforms stood out so clearly that the German soldiers, 1,500 yards away, were almost dazzled by the sight. The fact that the French also unfurled a regimental standard and had their band marching and playing alongside rendered them virtually impossible to miss and ensured that the regiment was almost annihilated. To add insult to injury the survivors were then hit by shells fired by their own artillery. This black month of August saw Méssimy's warning forcibly vindicated, the French suffering 206,515 casualties. Even then they did not learn. Elite regiments like the Zouaves and Spahis (infantry and cavalry regiments developed from France's colonial experience in Algeria) wore extraordinarily colourful uniforms, while the shiny breastplates and long horsehair plumes on Grecian helmets sported by the Cuirassiers provided useful targets for the enemy. Some other armies were equally obtuse: the Belgian cavalry of 1914 wore green tunics and purple breeches, while the Austrian horsemen had yellow breeches.

In 1914 officers of the Austro-Hungarian army were easily identified by their yellow sashes and their bright sword scabbards. As a result they became the main targets for both Serbian and Russian riflemen and in the first four months of the war 3,168 officers were killed and 7,781 wounded, a third of all the officers in the army.

Uniformly uncomfortable

Colour was one thing, but what about comfort? It was during the eighteenth century that rigid military discipline reached a high point and uniforms an impracticality that has never been equalled. The breeches worn by the eighteenth-century soldier were so tight that it became almost impossible to bend and an upright carriage and mechanical way of walking was inevitable. The tall, mitre-shaped hats of the Hessian infantrymen in the American War of Independence were constantly being knocked off by overhanging branches as well as providing American sharpshooters with unmissable targets. The stock – the leather collar almost universally worn in the eighteenth century – kept the head upright but was uncomfortable because it cut into the neck and chin. Gaiters took ages to do up as they had so many rows of tiny buttons. The hair on the head was pulled back in a queue (or pigtail), sometimes so tightly that a man could not even close his eyes. Regulations required a certain number of locks of hair on either side of the head and men were known to stay up all night preparing for the next day's parade. Such uncomfortable and constricting attire can hardly have been conducive to good performance on the battlefield. Close attention to sartorial detail may have been justified on European battlefields as an adjunct to perfect discipline, but when these troops were transferred to the wild forests of North America their uniforms became a positive drawback. The French were quick to adapt, the English much slower. The scarlet jackets of Braddock's infantry on the Monongahela provided the Indians with easy targets (see p. 86). Even when some enterprising redcoats broke ranks and took cover behind tree trunks or fallen logs they were beaten back into line by their officers, to stand and die in the European fashion.

When times were hard for Frederick the Great and the price of cloth was high, the Prussian soldiers found their coats cut from the minimum amount of material. This made the jackets so tight that the lapels could not be buttoned across in cold weather. In the damp the uniforms shrank and constricted the men's circulation unbearably. Early in the Seven Years War Frederick decided that he could not afford to equip his grossly swollen armies with cloaks and so the men froze in winter.

Boots, boots, boots . . .

Napoleon may have said that an army marches on its stomach, but as the Confederate soldiers of 'marching-generals' like 'Stonewall' Jackson knew, a good pair of boots was as important as a rifle. Some historians have gone so far as to argue that the defeat of the Confederacy owed much to the failure to supply their troops with adequate footwear. It is extraordinary to learn that Governor Vance of North Carolina hoarded 92,000 uniforms plus vast quantities of blankets and shoe leather at a time when troops from other Confederate states went short. North Carolina's 42 textile factories gave her twice the output of the rest of the Confederacy put together, and yet her belief in states' rights meant that she could see no further than supplying her own 'sons'. By the time the battle of Shiloh was fought in 1862, 60 per cent of Confederate soldiers present were wearing captured Union uniforms, hats and coats, with the obvious danger of casualties caused by mistaken identity.

But the main problem continued to be the lack of shoes. The strategy of Robert E. Lee depended on the fast flanking marches of his fellow commanders such as Jackson, A. P. Hill and Longstreet. This ruined shoes and boots at a fast rate and although fresh footwear was ordered from Europe as early as the First Battle of Bull Run in 1861, there were shortages of 40,000 pairs within months. Many Confederate soldiers had to be left behind on the marches or excluded from the battlefield for lack of shoes. This was so severe at Antietam in 1862 that it has been suggested that lack of shoes – and consequent lack of men – cost Lee a victory. The authorities had to take 2,000 men out of the army to concentrate on nothing but making shoes, yet at the battle of Gettysburg in 1863 thousands of Rebels – lacking suitable footwear – were still unable to take part in the battle. Hood's troops in Tennessee were particularly badly hit; at one point 25 per cent of them were obliged to go barefoot in frozen weather, snow and sleet. So acute were shortages that to sleep with your shoes off was to have them stolen. The Union troops, on the other hand, suffered not from a shortage of shoes but from poor-quality ones. Northern contractors supplied shoes of uniform shape for either foot, soldiers being supposed to wear them in. This caused much foot-soreness and straggling.

During the Crimean War there was a great scandal about the supply of boots to the British army. In the harsh conditions of the Crimean winter, men generally wore more than one pair of socks, and as the damp weather caused their feet to swell few of the available boots fitted. The Commissariat supplied boots of certain specified sizes, but paid no attention to the fact that in damp conditions they might need more of the larger sizes, and men frequently had to wear boots so small that 'women could scarcely have got them on'. British soldiers, who had viewed with horror the stripping of corpses and grave robbing by their Turkish allies, began to lose their qualms as their own clothes disintegrated. Midshipman (later Sir Evelyn) Wood paid a sailor ten shillings to find him some Russian boots from a graveyard. He was very pleased with those he got. Many British officers preferred captured Russian boots to the British ones, which were of such shamefully bad quality that 'the soles dropped off after a week's wear'. The inadequacy of the British boots was a result of contractors economizing to keep costs low, providing boots totally unfit for wet, muddy conditions. On 1 February, the 55th Regiment, parading in 'one vast black dreary wilderness of mud', sank into the slime, and the strain of pulling their feet out caused all the boot soles to be sucked off. The men apparently threw the boots away and marched to the front in their socks.

In 1914 the Russian troops often lacked boots, marching barefoot through the mud of autumn or wearing wooden shoes. Even as late as World War II the problem of soldiers' footwear had not been satisfactorily solved. During Operation Barbarossa in 1941 (see p. 173), Italian contingents with the German army had been equipped with cardboard shoes, even though expensive luxury leather shoes were still on sale in Italian shops. What happened to such shoes in a Russian winter can easily be imagined. Even the Germans found that their boots were not the equal of those of their Soviet opponents. The following anecdote shows how desperate and resourceful the Germans became:

> Now we had an opportunity to equip our men with more winter clothing. Kageneck ordered that the 73 dead Russians be carried to the village and stripped of their felt-lined boots and warm clothing.
>
> But the bodies were frozen stiff. And those invaluable boots were frozen to the Russians' legs.
>
> 'Saw their legs off', ordered Kageneck.
>
> The men hacked off the dead men's legs below the knee and put legs, with boots still attached, into the ovens. Within ten or fifteen minutes the legs were sufficiently thawed for the soldiers to strip off the vital boots.

Pack up your troubles . . .

The failure of the Commissariat to provide the British soldiers with adequate clothing and blankets in the Crimea was yet another scandal of that most desperate conflict. Problems had begun as soon as the British troops landed in the Crimea. Everyone was in such a hurry to advance on Sebastopol that the men were told to leave their knapsacks in the transport ships. British soldiers usually carried a change of clothing in their knapsacks and when the transports promptly sailed off – not to reappear for six weeks – the men were left with only the clothes they stood up in. When the transports returned the knapsacks that were still aboard had been ransacked.

In the harsh conditions of trench life in the siege lines around Sebastopol the clothes that the ordinary soldiers were wearing soon wore out. Without spare kit the injured men arrived at the hospital at Scutari in a wretched state, caked in mud and infected with lice (see p. 136). In wet weather the men had to wear their sodden kit day after day, sleeping wrapped in greatcoats and a single blanket on the damp and muddy floors of their tents, with nothing to prevent the effects of rising damp.

When news reached Britain of the destitute state of the troops, great quantities of warm clothing were rushed out. But the ship that was carrying them – the *Prince* – sank on 14 November, taking with it 40,000 greatcoats and boots for almost the entire army. As a result the men suffered appallingly during the month of December. Ironically supplies of 'rugs' came in by each ship until they numbered some 25,000 by January, yet in spite of the fact that they were as useful as blankets for keeping out the cold and damp, only 800 were supplied to the troops and no one had the imagination or initiative to issue any more. Of the thousands of palliasses shipped to the Crimea to protect the soldiers from the moisture of the ground, none were issued to the troops for lack of straw and hay with which to stuff them.

By the end of November 1854, 12,000 greatcoats had arrived safely at Balaclava, but, at the time when they were most desperately needed – the months of December and January, when men were dying of exposure – over 9,000 remained in store because the service regulation, established by Queen's warrant, said that soldiers should not be issued with greatcoats more than once in three years. Those who had lost them, on the march or in battle, would just have to wait. And so, in the last analysis, men died not from Russian bullets, or from disease, or from the freezing winter weather, but strangled by bureaucratic red-tape.

Florescu's flair

In the history of the Romanian Army the name of General Ion Emaoil Florescu must have a special place. Where other generals might have an eye for a defensive position or an advantageous geological feature, Florescu had an eye for haute couture, which would have done credit to a Parisian designer planning his Spring Collection.

In the period before World War I Florescu was obsessed with uniforms, insignia and medals, and went to incredible lengths to obtain the Légion d'Honneur for himself, not through some act of military daring for the benefit of the French nation, but by pestering French ambassadors at social functions. To Florescu an army had to have glitter because the uniform reflected the importance of the soldier in society. While he commanded the Romanian army there was a veritable orgy of regimental standards, emblems and commemorative plaques for troops to attach to their uniforms. Florescu actually took over the design of uniforms himself, with greatcoats, epaulettes and headdresses all receiving his own personal stamp of approval. Codes of dress were issued along with Orders of the Day, so that Florescu's soldiers became mannequins rather than fighting men. Special uniforms were created for engineers, brewers, baggage handlers and every other kind of ancillary worker who supported the troops. Make-up became common among the senior officers, though, ever conscious of social differentials, Florescu did not permit troops and NCOs to use it.

Florescu's became the gaudiest army in Europe and probably the least competent. When Romania entered the First World War on the side of the Entente Powers, Russian troops became frustrated at the number of times that huge numbers of well-dressed Romanians kept trying to surrender to them, mistaking them for the enemy. Having received directions from the Russians, they then set off looking for Germans and Austrians to take them prisoner.

Empty Stomachs

War is one of the most arduous activities undertaken by mankind and the physical condition of the men required to carry out the fighting, marching, entrenching and labouring is of paramount importance. Without adequate food the efficiency of soldiers will decline and their health will suffer. This seems almost too obvious to need saying, yet during the Crimean War men were exposed to some of the most difficult conditions ever faced by British soldiers, and supplied with a quite inadequate diet.

A convict's rations

In March 1855, Sir John McNeill and Colonel Alexander Tulloch were sent to look into the problem of feeding the British army in the Crimea. Their report shocked the British people out of their complacency. When they visited the hospitals at Scutari they were told that sick men arriving there were suffering from diseases attributable to poor diet. Compared to the French, dietary arrangements for British troops were primitive. During the winter, they learned, British soldiers mainly ate salt meat and biscuit, with a 'very insufficient proportion of vegetables'. Yet it was common knowledge that the men's health would suffer without supplies of fresh meat, vegetables and freshly baked bread.

The Commissioners reached the conclusion that 'deaths . . . amount to about 35 per cent of the average strength of the army present in the Crimea from 1 October, 1854, to 30 April, 1855, and . . . this excessive mortality is not to be attributed to anything peculiarly unfavourable in the climate, but to overwork, exposure to wet and cold, improper food, insufficient clothing during part of the winter, and insufficient shelter from inclement weather . . .' Something had to be done – and quickly.

The Commissioners were concerned with both the quantity and the quality of the soldiers' diet. Men often only got half rations and sometimes no food at all. On Christmas Day 1854, Colonel Bell's men received no rations. The Colonel then 'kicked up a dust' with a Commissary officer and eventually some small portions of fresh meat were served out, but by that time it was dark and the men had no fires or other means of cooking it. In any case, salt meat and biscuit did not supply enough nutritional value to men working such long hours and in such conditions. Many soldiers could not even eat salt beef because it gave them diarrhoea. So a lot of it was thrown away, hundreds of pounds a day by some regiments. Even when fresh meat was available the men were given salt meat instead because the Commissary General claimed it was too much trouble to issue.

Past experience had shown that too much salt meat increased the risk of scurvy. However, because it was more trouble for the Commissariat to issue fresh meat than salt meat, and to bake fresh bread rather than distribute bags of biscuits, thousands suffered unnecessarily. Colonel Tulloch compared the weight and nutritional value of the daily rations of a British sailor, a Hessian soldier and a British soldier in the Crimea. The nutritional weight of the sailor's rations was 28.5 ounces, the Hessian's 32.96 ounces, while the Crimean soldier's was a mere 23.52 ounces. In addition, the soldier's salt meat was less nutritious than fresh meat and in the Crimea he mainly threw it away because of the irritation it caused his bowels, surviving on rum and biscuit instead. It is a wonder that anyone survived at all. Even an inmate of a Scottish prison, with no exposure to bad weather, received 25.16 ounces, including bread, vegetables, fish and milk, so the British soldier was worse off than a convict.

The Commissariat seemed to get everything wrong. On 4 November, 1854, 150 tons of vegetables were shipped on board the *Harbinger*. But the ship left the Bosphorus without the correct papers and when it arrived at Balaclava its captain could not get anyone to accept responsibility for the cargo. While officials squabbled over chits and dockets the food rotted, eventually being thrown overboard or eaten by roaming bands of French Zouaves. None of these vegetables reached the British troops at the front. From November to March there was not enough fresh meat to maintain the health of the army, while the lack of vegetables was so severe that most men had a monthly ration of only two potatoes and one onion. When the army was in Bulgaria, Lord Raglan had authorized the daily issue of two ounces of rice to each man, as so many men had bowel complaints. But when this order lapsed on 15 November, it was not renewed because nobody remembered to do so. The Commissariat claimed that though they had rice they had no means of getting it to the troops. Why the men could not fetch it themselves is difficult to understand, particularly as this was the only vegetable matter available and vital in combating scurvy.

Had Lord Raglan but known it there were available at Balaclava and Scutari rice, potatoes, peas and Scotch barley, which would have been far more beneficial to the men's health and no heavier to carry than salt meat and biscuit. But the Commissariat had made it clear to Lord Raglan that it was not their job to supply vegetables because 'according to the usage of the service' it was the men's own responsibility.

Scurvy

One of the worst failures of Commissary-General Filder concerned the issue of lime juice. It had been known for a century that the scurvy which affected ships' crews on long journeys could be prevented by the issue of citrus fruit. This was a well-established principle by the time of the Crimean War and the fact that so many British soldiers succumbed to scurvy was nothing short of a national outrage. On 10 December 1854, the *Esk* reached Balaclava harbour carrying 278 cases containing nearly 20,000 pounds of lime juice, to be issued generally to the troops. In the previous month small quantities had been obtained from naval sources to help cure scurvy among the sick, but no general action had been taken to counter the rapid spread of the disease through the British camp. Yet from 10 December to the first week in February, the cargo of lime juice remained untouched on board the *Esk*, even though Filder knew it was there, because he claimed it was not his job to tell the army that it had arrived.

McNeill and Tulloch reported that until April 1855 the troops had no bread and had to make do with hard biscuit. This was particularly tough for those with scurvy, who could only eat the biscuit in great pain from their inflamed gums. Bread was apparently available from private sources in Balaclava, but at a price far beyond the ordinary soldier's pocket. The French had established bakeries and were supplying their own soldiers with fresh bread. The British could easily have done this if anyone had taken the trouble to arrange it – there were enough bakers in the regiments to make it a simple task.

Green coffee

Of all the problems facing British soldiers in the Crimea it was the issue of green coffee which caused the most irritation. This was the fault of Commissary-General Filder, who asked that the coffee beans should be sent to the Crimea unroasted, because they were less prone to dampness and mould on the journey. The result was predictable. Without any way of roasting and grinding the beans the soldiers often drank a foul concoction which was actually harmful to them, according to their medical officers. The commander of the 1st Regiment commented:

> A ration of green raw coffee berry was served out, a mockery in the midst of all this misery. Nothing to roast coffee, nothing to grind it, no fire no sugar; and unless it was meant that we eat it as horses do barley, I don't see what use the men could make of it, except what they have just done, pitched it into the mud!

One can hardly blame the men for lack of enterprise, some finding cannon-balls and empty shell cases to grind the coffee berries, while others cut up their dried meat into strips and used it as fuel to roast their coffee. Ironically, in the stores at Balaclava there were some 2,705 pounds of tea which had been forgotten and were not issued.

Blame your Tools

A soldier or airman has a right to know that the weapons he takes into battle with him are likely to prove efficient when put to the test. His own life depends on him being able to defend himself with such weapons and the success of his mission on their correct functioning in the face of the enemy. But the weapons in which soldiers have placed their trust have not always proved reliable on the field of battle.

Bombs, tanks and muskets

The low level of reliability of the bombs dropped by British aircraft during World War II meant that night after night over Germany half the crews of British bombers were risking their lives for nothing. They might as well have dropped baked bean cans. Official reports acknowledge that a third of all medium-capacity bombs broke up on impact and failed to detonate, while the picture was even worse in the case of high-capacity bombs or 'blockbusters'. If one adds the fact that only one bomber in ten actually got to within five miles of its target, that its bombs when dropped fell randomly over an area of 75 square miles, and then in 50 per cent of cases failed to explode, one is obliged to call into question the competence of Bomber Command. It is a little known fact that during the bombing raids of 1940–1, more British fliers died than German civilians.

'Are you aware it is private property? Why you will be asking me to bomb Essen next.'

Sir Kinsley Wood, Secretary for Air, September 1939, on a plan to burn the Black Forest

British attitudes towards total war in 1939 were pitifully naïve. While Hitler crushed Poland the RAF confined itself to dropping leaflets on Germany.

Before politicians push a crisis to the brink of war they should first make certain that they have the men and weapons necessary to prosecute their cause. In 1956 the British Conservative government of Sir Anthony Eden decided to take military action in response to the nationalization of the Suez Canal by President Nasser of Egypt. But Eden's belief that Britain was equipped for an amphibious landing in the canal zone was gravely mistaken. Many British weapons were outdated or of doubtful use in desert conditions. The BAT anti-tank gun had only been tested in tropical conditions and since it was felt that it might malfunction in a sandy atmosphere it had to be left behind. The FN self-loading rifles were also felt to be questionable and so the infantry had to use rifles of early World War II vintage. For a sea power supposedly specialized in amphibious warfare, it is shocking to learn that Britain possessed hardly any assault landing craft or tank landing craft. Some of those kept in 'mothballs' for such an emergency were found on inspection to be rotten, while others were doing duty as pleasure steamers or ferries. To add a final touch of farce to British preparations, it was found that the army did not have enough tank transporters to move the Centurion tanks to Southampton.

Thus most of the tanks arrived in Southampton courtesy of Pickford's Removals, in that company's giant pantechnicons. It took four weeks to transport 93 tanks from Tidworth to Southampton.

The Egyptian tanks in the 1967 war against Israel had been bought from the Soviet Union. Unfortunately, they had not been personalized, and in the broiling desert temperatures the Egyptian tank crews found that their tanks had no air conditioning, only heaters suitable for Russian winters.

The Prussian army of Frederick the Great, so famous for its iron discipline and perfect drill, was equipped with some of the most inefficient weapons of the eighteenth century. The infantry musket was a particular offender. Its long barrel (41.5 inches), bayonet and cylindrical ramrod rendered it muzzle-heavy to the tune of three pounds. With such a heavy weight at the muzzle end there was an inevitable tendency to shoot low, and Prussian marksmanship, accordingly, was very poor. Even in training, few men could hit anything other than the bases of wooden targets, and most shot into the ground. In addition, during drill stocks were constantly banged into the ground, causing them to split, and so much polishing of the iron barrel was carried out – for display purposes – that the metal became worn away. Nor were officers and NCOs any more fortunate, having to carry spontoons – a kind of pike eight feet in length – which were simply a hindrance and were often carried by officers' servants instead. Even worse were the NCOs' ten-foot half-pikes, so heavy that few could even carry them.

The needle gun

Rarely has the difference between appropriate and inappropriate weapons been so clearly demonstrated as at the battle of Königgrätz in 1866. Here the Prussian needle gun won a victory over the Austrian Lorenz rifle that Prussian tactics hardly deserved.

'Although the needle gun permits rapid fire as long as there is no stoppage, this does not constitute any real advantage, because rapid fire will merely exhaust the ammunition supply.'

Feldzeugmeister Augustin, 1851

The Austrians rejected the quick-firing needle gun. The Prussians adopted it by 1866. At the battle of Königgrätz it was the needle gun that made the difference between the two sides.

The Austrians had had the opportunity of equipping their own troops with needle guns, having tested them in 1851 and rejected them as not being cost efficient. They claimed that the speed of firing that would be possible might make soldiers waste ammunition. In addition, the Emperor Franz Josef, after his misfortunes at Magenta and Solferino against the French in 1859, had reached the conclusion that the bayonet was the king of weapons. As a result, the Austrians adopted what was known as the *Stosstaktik*, which emphasized all-out attack and coming to grips with the enemy, rather than concentrating on long-range marksmanship or concentrated firepower. During the

Austro-Prussian war against Denmark in 1864, the Austrians adopted this principle and won the admiration of their then Prussian allies for their bravery and willingness to endure high casualties. The Austrians were to pay a heavy price for these tactics when facing the quick-firing needle gun at Königrätz.

There was another reason – and one far less forgivable than the antique military theories of an absolute monarch – for the Austrians being sent into battle with an inferior weapon. The Vienna arsenal, which made the Lorenz rifles, had a monopoly of Austrian weapons manufacture. Having just introduced expensive new machinery to increase production of the Lorenz to 1,000 rifles a day, it is not hard to imagine the reaction to the idea of going over to a new rifle. Only in the last few weeks before war in 1866 did the Austrian military authorities come to their senses. Knowing that their men would be outgunned by the Prussians, they made desperate efforts to buy needle guns abroad. Approaches were made to the United States and Belgium, but the first consignment of 5,000 modern Remington rifles only arrived as the war was coming to a close.

The Austro-Prussian war of 1866 opened with an Austrian success at Trautenau. Yet even this victory for the *Stosstaktik* showed its essential flaw: Austrian casualties in victory were far higher than those of the Prussians in defeat. When the main armies clashed at Königrätz it was too late to change anything. The triumph of the needle gun meant that Austria suffered defeat and 40,000 casualties, while Prussia won not just a battle but supremacy in Germany.

Impressive follies

During the Renaissance military technology often entered the realms of fantasy. Artists, sculptors and inventors revelled in the opportunities offered by sixteenth century military commanders, who demanded ever more sophisticated – and impractical – means of destruction. Antonio della Scala produced one amazing weapon – the *Ribaudo* – a kind of machine gun with 144 barrels loaded with shot, of which any 12 could fire simultaneously. It might have proved a deadly weapon but it was so heavy that it needed four sturdy horses to pull it. There is no evidence that it was ever fired because there was never time to get it in position to do so. In its impracticality it resembled the extraordinary warships of the Hellenistic period, developing from bireme, through trireme to quinquireme, to contraptions of such weight and complexity as to beggar description. The larger vessels, with multiple decks and oars so numerous they resembled the legs of some nightmare insect, sometimes turned over or sank on launching due to their enormous displacement.

Surprises in store

Lack of ammunition can be as fatal as lack of correct weapons. Three examples from British colonial campaigns show just what can happen if insufficient attention is paid to ammunition supplies.

Charles Brandon, the manager of British military stores at Cape Coast in West Africa in 1824, bears a heavy responsibility for the fate of a British force led by Sir Charles

Macarthy. Macarthy's men were under attack near the village of Bonsaso from an army of 10,000 fierce Asante tribesmen. Macarthy had tried playing 'God save the King' to repel them, but this had not of itself held them back. The British redcoats were heavily outnumbered and had formed a square to hold back the attacks of the Asante warriors. In the heat and dust of battle the soldiers began to run short of ammunition and Macarthy ordered Brandon to break open the reserve ammunition boxes, which he had recently brought up from the coast. As the Asante closed in for the kill, the ammunition boxes were smashed open only to reveal that they were full of biscuits! Macarthy ordered his men to retreat, but his camp was overrun by the Asante, who killed most of the redcoats, including the unfortunate general, whose head was cut off and displayed on a spear. The next time his skull was seen it was being used as a drinking cup by the Asante chief.

During the Gordon Relief Expedition in the Sudan in 1885, the Dervishes attacked a British force at Abu Klea, achieving the unthinkable by breaking the British square. During the fighting the reserve ammunition boxes were opened, one of which was found to be full of gold sovereigns intended as payment to friendly tribesmen. Even though the battle was raging around them some men were overcome by greed and tried to stuff their pockets with the coins instead of Lee Metford rifle cartridges.

At the battle of Isandhlwana fought between British and Zulu forces in 1879 the red-coats had started the battle with only 40–50 cartridges each and so heavy was the firing that shortages soon became acute. Officers were not worried at first, knowing that there were ample reserves in the wagons, and men were sent running back to bring fresh car-tridges to the firing line. The regimental reserve ammunition was packed in heavy wooden boxes, held down by copper bands and nine screws each. The two quarter-masters each had screwdrivers, but they found that some of the screws had rusted and

The aftermath of the Battle of Isandhlwana during the Zulu War of 1879. The British were caught unawares by a Zulu force of 20,000, and 1,600 were killed in hand-to-hand fighting. A failure of reconnaissance was compounded by a breakdown in ammunition supply, turning defeat into disaster.

were difficult to release. Also, regulations would not allow them to open more than one box at a time, as every cartridge had to be accounted for. When the Natal Native Horse sent for ammunition they were refused it by the quartermasters of the 24th Regiment and had to try to find some elsewhere in the camp.

Lieutenant Smith-Dorrien, later a World War I corps commander, noticed the British fire slackening all around and collected some men to try to open several other crates. But there were no spare screwdrivers. They hacked at the boxes with axes or stabbed them with bayonets, splintering the wood. Forcing open the top Smith-Dorrien began shovelling cartridges into men's helmets only for one of the regimental quartermasters to tell him to stop as he had not got the right requisition papers. While the Zulus picked off redcoats, red tape still ruled within the ammunition wagons. The British fire had almost stopped and men were running back from the firing line, desperate for cartridges. Twenty thousand Zulus now drove in the advanced cavalry units and entered the camp. Panic spread and organized resistance collapsed. The British camp was overrun by the Zulus and most of the troops massacred.

Sick Parade

Throughout history disease has caused far more casualties in wartime than battle. Wherever large numbers of men are brought together in poor conditions and made to undergo arduous physical exercise, often with inadequate food and clothing, disease has been ever present. However, in many cases commanders have failed to take the measures available to them to avoid losses caused by ill health. Many problems have stemmed from simple neglect, as when thousands of British and French soldiers serving in the West Indies during the eighteenth century succumbed to yellow fever. Despite the improvement of medical knowledge in the nineteenth century, the British treatment of wounded and sick men during the Crimean War was woefully inadequate. Medical services during the Mesopotamian campaign in the World War I were little short of medieval. Just as the commanders who 'butchered' their men in battle stand accused of military incompetence so do those whose neglect of basic medical precautions imposed suffering without need.

The Age of Enlightenment

Medical services in the Prussian army of Frederick the Great were virtually non-existent. Once a man was wounded he became merely a burden on the army and warranted little attention. After the battle of Hochkirch against the Austrians in 1758 the wounded Prussians were not provided with special transport and had to find their way back to the field hospitals as best they could. Walking wounded piled the more serious cases on to carts and waited for them to die so they could steal their clothes. Food was not allocated to these injured soldiers and they were forced to beg in the villages that they passed through. When the injured reached hospital they were always in danger of losing their clothes and possessions to the medical orderlies. The head of the hospital, being a serving officer without medical knowledge, regarded his role as getting the slightly wounded back into the ranks of the army and neglecting those who were of no further use. A Prussian officer described his own experiences in 1757 at a hospital in Dresden:

> I was horrified when I entered the house of misery and saw a great quantity of corpses heaped on top of each other along the walls, with their rigid feet poking through the straw which had been scattered over them . . . Whole rooms were filled with sick of every category, who were stacked up in layers and afforded the most inadequate treatment.

During the War of the Bavarian Succession in 1778, only one man out of every five who entered a Prussian hospital for treatment came out alive. After the battle of Torgau in 1760, an observer commented:

> The cold kills off most of the wounded, as usually happened in the Prussian service, where the hospitals are so badly served and so malodorous that the soldier considers himself dead once he enters the portals. It is not altogether surprising that there are so few disabled men to be seen in the states of the King of Prussia after such a cruel war. I have it on good authority that the hospital directors and surgeons were under orders to let men die if they are wounded in such a way that they would be incapable of serving after they were healed.

The Grand Army

Under Baron Larrey, the medical services in Napoleon's army were far better than those of any other European army of the time. However, even though the plans for the invasion of Russia in 1812 had been more carefully drawn than for any earlier campaign, Napoleon and his advisers overlooked the vital fact that typhus was endemic in Poland and Russia. Typhus, sometimes called 'gaolbird fever', is a campaign disease spread by the lice which flourish in the unwholesome conditions of military camps. The areas of Poland through which the French army was to pass were unusually filthy even by the standards of the age. Before 1812 there had been no signs of a typhus epidemic in the French army, but in the abnormally hot, dry summer of that year the wells in Poland became contaminated and the water thick with organic matter. With the army under constant harassment, the supply wagons were forced to travel at the rear of the columns for protection, which meant that food was slow to reach the forward troops and sometimes never arrived. Napoleon's Grand Army had outgrown a command structure that was essentially of the eighteenth century in its organization.

As the French army crossed the River Niemann typhus began to appear, with men becoming ill with high temperatures, blotchy pink rashes and faces of a dusky blue tinge. By late July, 80,000 men had died or were sick with the disease. On the advice of his marshals, Napoleon called a two-day halt at Smolensk, but then changed his mind and pushed on towards Moscow. This was a blunder of the first order. Disease was inflicting heavier casualties on the army than would have occurred in several hard-fought battles. By September, Napoleon's army had shrunk to 160,000 effective men, and two weeks later to 130,000. By the time he entered Moscow he had just 95,000 fit soldiers. While summer heat and the typhus of Poland and Russia had defeated Napoleon even before the battle of Borodino, the bitter Russian winter would now transform defeat into catastrophe. Of Napoleon's original fighting force of 450,000

Despite not losing a battle during the campaign, Napoleon's invasion of Russia in 1812 constituted his greatest defeat. The bitter Russian winter combined with the inadequacies of the French logistical system to take a terrible toll: only 40,000 of the original French force of 450,000 men returned.

men, no more than 10,000 would ever be fit enough to bear arms again. Without losing a battle, the Emperor had suffered one of the greatest defeats in human history.

Crimean catastrophe

The Crimean War has deservedly become a byword for the neglect of welfare services. A single glimpse at conditions in the Barracks Hospital at Scutari as Florence Nightingale found it serves to indict Britain's military leaders for their incompetence.

The attitude of the medical authorities in the Crimea can be summed up by Mr Ward, Purveyor at Scutari and a veteran (aged 70) of the Walcheren expedition during the Napoleonic Wars (see p. 145). When questioned by the Hospitals Commission in December 1854 about conditions at the Barracks Hospital, he replied:

> I served through the whole of the Peninsular War. The patients never were nearly so comfortable as they are here . . . In general, the men were without bedsteads. Even when we returned to our own country from Walcheren and Corunna the comforts they got were by no means equal to what they have here.

In fact, what Ward considered 'comfortable' was little more than an open sewer running in filth. The whole hospital thronged with wounded men, who sometimes languished for two weeks before being seen by a doctor. The floors on which the men lay were unwashed and home to vermin and insect life. When a visiting chaplain tried to talk to the wounded men he got covered in lice. Without pillows, sheets or blankets, the men

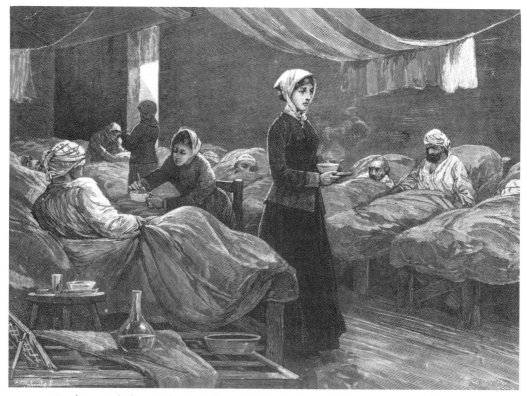

Conditions in the barracks hospital at Scutari during the Crimean War were nightmarishly squalid.
Florence Nightingale drew the scandal to the attention of the British public on her arrival there in
December 1854.

rested their heads on their boots and covered themselves with their greatcoats, frequently caked in mud and blood. Operations were carried out in the open ward and the screams of the men having limbs amputated must have terrified those whose turn was yet to come. Florence Nightingale at least succeeded in having these scenes screened off, but nothing could contain the noise.

Miss Nightingale estimated that during her visit there were as many as a thousand men in the hospital with gastric problems, notably diarrhoea and dysentery. Yet there were just twenty chamber pots to go round, and the latrines lacked running water to flush away the filth. So foul had conditions become that the contents of the privies overflowed and covered the floor of parts of the hospital to a depth of an inch or more. Without shoes or slippers the wounded men were forced to paddle through this every time they answered the call of nature. To avoid this horror, large wooden tubs were placed in wards and corridors for the men to use, but these soon polluted the whole atmosphere of the hospital. In conditions as insanitary as these, men entering the hospital with simple wounds soon succumbed to an appalling list of complaints. One in every two men admitted died of dysentery, and wounds rapidly became gangrenous and necessitated amputation – itself a death sentence in most cases.

After the battle of the Alma, in September 1854, Surgeon George Lawson wrote:

On the 20th, when the Light Division had about 1,000 killed or wounded, there were no ambulances, etc., or lights (save the personal property of the officers) –

nearly all the operations requiring to be performed on the ground. I, myself, oper- ated the whole of the first day on the poor fellows on the ground, and had per- formed many on the second, until an old door was discovered, of which we made a table . . .

The callousness of the medical orderlies has to be judged by the standards of the time and the exigencies of wartime. Yet reports which speak of Balaclava harbour filling with amputated arms and legs in the water, or limbs piling up outside hospitals to be eaten by the hogs, strike the twentieth-century reader with a true sense of horror.

Self-inflicted wounds

The high rate of mortality among officers in the Prussian army in the eighteenth century may have been because many of them were so seriously affected by venereal disease that even minor wounds incapacitated or even killed them. The example of Field Marshal Seydlitz can be cited in support of this idea. He was wounded both at Rossbach in 1757 and at Kundersdorf in 1759, and on both occasions his recovery was impaired by the fact that he was suffering from syphilis.

The consequences of venereal disease – the most avoidable mishap that can affect a soldier – have been enormous throughout history, although statistics have only been readily available in the present century. For the entire British army during World War I, 416,891 men were admitted to hospital suffering from VD – a quarter of all non-combat-related admissions. If one considers that the average time spent in hospital was over six weeks per patient then the drain on manpower can easily be imagined. In 1914 the Russian authorities miscalculated badly in their provision for the sick and wounded, finding that the number of hospital beds available for battle casualties sufficed only to house the number of their soldiers suffering from venereal disease.

In the interests of military efficiency the authorities have always tried to prevent sol- diers from contracting venereal diseases. On the basis that it is impossible to remove the sexual urge from millions of young men, some concluded that the only alternative was to provide rigorously controlled outlets in the shape of regulated brothels, where prosti- tutes could be medically examined. The French were more progressive in this matter than the Anglo-Saxon nations and were providing brothels for their soldiers from as early as the 1840s. During the war in Indo-China in the late 1940s, two prostitutes were awarded the *Croix de Guerre* for their services to an isolated French garrison. In 1918 President Clemenceau of France even offered to help the Americans organize licensed brothels, but his suggestion was turned down by the Secretary for War, on the grounds that if President Wilson ever got to hear about it he might even withdraw US military assistance from the Allies.

The British Army in India was notably prone to venereal infection. During the 1830s there was an annual figure of over 300 men per 1,000 suffering from venereal diseases. Throughout the century a continual battle was fought between those who favoured reg- ulating brothels for soldiers, thereby reducing casualties from venereal infection, and others who felt that to do so was to encourage the men in acts of lewdness and inde- cency. When, in 1870, the Quartermaster General issued instructions to British officers to ensure the effective inspection of regimental prostitutes, there was an immediate out- cry in Britain on the grounds that 'no considerations of health, economy or expediency can justify or excuse legal prostitution'. Figures for admissions to hospital for VD cases increased even more, reaching 361 per 1,000 in 1887 and 370 the next year. Anglican bishops were clear in their view that no argument on sanitary or health grounds could be acceptable in condoning vice. By 1890 the rate was approaching 50 per cent of the white strength of the Indian Army. Many men were so badly affected that their army careers were ended and they had to be sent home to Britain for further treatment. In the period 1880–1900 over 10,000 British soldiers were invalided home with severe venereal infection, sometimes leading to permanent disablement and death. In 1903 General Kitchener tried to dissuade soldiers from consorting with Indian women, warning them that venereal disease 'assumes a horrible loathsome form . . . the sufferer finds his hair falling off, his skin and the flesh of his body rot, and are eaten away by slow cancerous ulcerations, his nose falls off, and he eventually becomes blind . . . his throat is eaten away by foetid ulcerations which cause his breath to stink'. But the sexual alternatives, masturbation or homosexuality, were even worse in the eyes of Victorian society. The former was thought to lead to blindness and insanity, the latter resulted in social ostracism.

During World War II the British authorities bowed to the inevitable and allowed brothels to remain open in military areas, provided the girls submitted to regular medical check-ups. In Tripoli separate brothels were opened for officers and men, though the Chaplain-General soon put a stop to the official brothels, leaving only the unregulated ones. In Cairo and later in northwest Europe, General Montgomery 'scored an own goal' by closing regulated brothels. The results were predictable and most unfortunate. As one army doctor reported: 'Within three weeks every bed in the previously almost deserted VD ward, and every bed that could be crammed on to the verandah outside, was full.'

On many fronts the number of cases of VD exceeded not only the incidence of any other disease, but even combat casualties. During the heavy fighting in the Middle East in 1941 battle casualties reached 35 per 1,000; in the same period VD casualties were 41 per 1,000. In Italy in 1945 battle casualties of 9 per 1,000 were matched by VD casualties of 68 per 1,000. In Burma in 1943 battle casualties of 13 were small compared to VD casualties of 157 per 1,000. The figure for black American soldiers in Tunisia in 1943 reached the extraordinary level of 451 per 1,000.

This was no way to run a war, particularly as simple solutions were at hand. Yet the American authorities were afraid to offend public opinion at home by offering condoms to the men. Subterfuge was the answer. When the Army Supply ordered a consignment of millions of condoms it was explained that the rubber contraceptives were used to keep moisture out of the barrels of machine guns and other automatic weapons.

Be Prepared . . .

It is one thing to arm a soldier with a weapon; it is quite another to train him in its use. Deficiencies in training have been widespread in military history, but with the mass armies of the nineteenth and twentieth centuries the problem has become very significant. In the American Civil War, the quality of soldiers used by both sides during the early battles was not high. At the battle of Bull Run in 1861 they were little better than armed mobs. Such enthusiastic amateurs needed to be turned into seasoned professionals, particularly as military technology was undergoing rapid changes. In 1861 the normal infantry weapon for both sides was the muzzle-loading, single-shot Minie rifled musket. It fired a 0.58-calibre bullet called a Minie ball, named after its French inventor. A marksman – and there were some in 1861 – could hit a bull's eye at 300 yards and penetrate six inches of pine at 500 yards. In the four years of fighting, the Unionists used nearly one and a half million of these muskets.

More haste . . .

In 1861, both Union and Confederate military planners needed to arm their men quickly and this meant buying from Europe whatever type of firearm was available. The most popular was the British Enfield rifled musket, though not every unit was lucky enough to get these. The Pennsylvania Reserves got instead 0.69-calibre smoothbore muskets of 1837 vintage, primitive weapons indeed. The Head of Army Ordnance in the North believed that giving multiple-shot, breech-loading weapons to undisciplined men would just encourage them to fire wildly and waste ammunition. If they had just one shot each time before they had to reload, they would think more carefully and not waste it. In fact, he was quite wrong. Muzzle loaders had to be loaded in a standing, hence dangerous, position, which caused men to fire quickly so that they could reload – again and again. Men armed with the breech loaders, having the confidence to take their time and really aim their weapons, scored better results every time. There was no fooling the men themselves, whatever the army experts said. Many soldiers saved their pay so that they could buy themselves a Henry (16-shot repeater) or Spencer (7-shot repeater).

Whatever the quality of the weapons, the competence of a regiment depended on the weapons-training they received. When troops are trained during wartime there are always pressures to get the men to the front before they are really ready. As a result training was patchy. At Shiloh in 1862, one 'green' regiment only discovered that they had not been shown how to use the weapons that had been issued to them after the battle had started. With fighting raging around them, veterans from another regiment had to be sent to give quick instructions. Even as late as Gettysburg in 1863, General Meade's officers collected 37,000 rifles that had been thrown away on the battlefield. More than a third had been loaded with more than one cartridge, some with as many as six. Few of the men had apparently realized that they needed to put a percussion cap under the hammer of the weapon, otherwise they could pull the trigger all day without it firing once. In the noise of battle they might not even have noticed that they were not firing. Meade found this proof of poor training methods sufficiently alarming to order immediate firing instructions throughout the Army of the Potomac.

. . . less speed

During World War I the Russian 'steamroller', which inspired both Austria and Germany with such terrors, often functioned with neither drive nor steam. Administrative failures led to many Russian soldiers being sent to the front without a rifle. Their instructions were to wait patiently in the trenches until one of their comrades was killed and then take their rifle. During 1915 sections of the Russian Twelfth Army had to be pulled out of the line because they did not know how to use their Japanese rifles. With many untutored peasants in their vast army the Russians suffered occasionally as a result of their own men's ignorance of the modern world. Russian aeroplanes were often shot at by their own side because the peasant conscripts believed such clever things as planes must have been made by Germans rather than Russians. Sometimes when the telegraph service broke down it was because the Russian soldiers, in their search for firewood, had chopped down the telegraph poles.

At the battle of Wilson's Creek in August 1861, General Siegel's German volunteers showed their lack of training. Having been successful with their initial attack, the famished Germans looted the Confederate camp and sat down to eat the breakfasts the routed Rebels had not had time to eat. Abandoning all restraint they refused to shift from their pork and beans until the Confederates, re-formed by their commanders, stormed back into the camp to regain their victuals.

Desertion

Desertion either before the battle or panic and flight during it have frequently been the result of poor training. Although it is an aspect of military history that receives little attention in regimental histories or memoirs of great commanders it is undoubtedly an important one, which has affected the outcome of more battles than most people care to admit.

During the Seven Years War a Prussian officer wrote: 'In my opinion the main cause of the prevailing and heavy desertion is to be traced to the fear of the soldiers that they will fall ill, and then be virtually buried alive in one of our overcrowded hospitals.' Whatever the cause, desertion from the Prussian army during the period 1756 to 1763 reached staggering proportions. The most prestigious regiment in Frederick the Great's army lost by desertion three officers, 93 NCOs, 32 musicians and 1,525 men, to add to the 130 who committed suicide and the 29 who were executed. From German sources we know that at the battle of Prague in 1757 many of Frederick's officers and NCOs, including one of his own aides, hid behind a hill during the fighting. In the same year, at Gross Jägersdorf, the bushes and undergrowth were alive with thousands of Russians who had decided to sit the battle out. At Torgau in 1760, non-participating Prussian and Austrian troops mingled happily together in the safety of the woods until the outcome of the battle was decided.

Lest it be thought that there was something unusual about the Seven Years War – an extremely bloody conflict – some examples from the American Civil War will show that this problem was far more widespread than historians have previously thought. At the battle of Malvern Hill in 1862 Confederate general Jubal Early met large numbers of men fleeing from the field, found others skulking in ditches and hundreds hiding in the woods. At Shiloh in the same year, a Tennessee regiment took fright during the advance and ran back into their camp shouting, 'Retreat! Retreat!' Having been rallied, they advanced again only to panic at the sound of gunfire and flee, trampling the regiment behind them and flattening its flag-bearer in the mud. The Texas regiment behind them panicked in turn and when their officers tried to stop them from fleeing they were fired at. The most extraordinary case of panic occurred at Missionary Ridge in 1863, when

Union troops scaled an almost sheer ridge and drove off superior Rebel troops. Confederate general Bragg wrote of the panic that ensued among his troops: 'It was a panic which I had never before witnessed which seemed to have seized upon officers and men, and each seemed to be struggling for his personal safety, regardless of his duty or his character.' At Winchester in 1864, Confederate general Stephen Ramseur commented:

> My men behaved shamefully. They ran from the enemy . . . The entire command stampeded. I tried in vain to rally them, and even after the Yankees were checked by a few men I posted behind a stone wall, they continued to run all the way to the breastworks at Winchester – and many of them threw away their guns and ran on to Newtown six miles beyond. They acted cowardly and I told them so.

Some men even exploited the dispensation that shoeless men need not fight by throwing away their boots, while others pretended to help the wounded or spent their time removing casualties from the field. The problem was an enormous one in both Union and Confederate armies. By 1863, up to 100,000 men on the Confederate side had avoided military service by deserting. In February 1865, 400 men from General Price's command deserted in a single day, and later in the same week a whole brigade deserted en masse. By the time the South surrendered the incredible figure of 55 per cent of their soldiers had deserted.

Maps

During the Crimean War there was such a shortage of adequate maps that one officer wrote home to his mother asking: 'Will you also be kind enough to send me a map of the Crimea with the forts, etc., well marked out in Sebastopol. I see them advertised at Wylds in the Strand. You can choose which you think best and send it by post.'

It is doubtful if the Russians were any better off themselves. Before the battle of Inkermann (see p. 51) the Russian commanders had no maps and were very vague about the positions they were to occupy. Colonel Popov, of Prince Menshikov's staff, whose job it was to see that maps were made available, tried instead to talk to the various commanders before the battle but did not manage to reach all of them. One commander, General Soimonov, got totally lost through ignorance of the terrain and the heavy fog.

Even Frederick the Great was careless about maps, preferring to keep his knowledge of terrain in his head. Unfortunately he had not devised a way of making his fellow commanders see things as he did. Before the battle of Kolin in 1757 he realized that he had no maps to issue. Assembling his generals he said that as many of them had been near Kolin in 1742, they ought to remember the lie of the land. 'I have a plan somewhere,' he admitted, 'but Major von Greise cannot find it.' At Kolin he was to suffer one of the worst defeats of his career at the hands of the Austrians under Marshal Daun.

Maps were a notable problem for commanders on both sides during the American Civil War. Nor should it be thought that a war fought on home ground made things any easier than a war fought in unfamiliar territory. In 1862 Union general Henry Halleck campaigned in the West with a map he bought from a bookshop, while the following year the Army of the Potomac still had not got accurate maps of North Virginia, even though they had been campaigning there since the outbreak of hostilities eighteen months earlier.

The English fleet before Cadiz. The first quarter of the 17th century saw a decline in England's naval strength. Some of the ships sent on the doomed expedition of 1625 were aging and sluggish survivors of the Armada and were ill-equipped for the demands of an ambitious campaign.

THE EXPEDITION TO CADIZ (1625)

The English sea dogs who had destroyed the might of Spain's Armada in 1588 would have turned in their graves if they had seen the expedition England sent against Spain in 1625. Compared to men like Drake and Hawkins, Lord High Admiral George Villiers, first Duke of Buckingham, was nothing more than an ambitious courtier. He knew nothing about the sea. Although he eventually handed over command of the expedition to a sound military man, Sir Edward Cecil, Buckingham appointed his friends and relations to all the other positions of trust. Significantly, of the six senior commanders not one had any experience of maritime warfare.

The fleet Villiers raised was worthy of the men who were to command it. There were altogether ninety vessels, including nine King's ships, supposedly heavily gunned, and about thirty armed merchantmen. But the rest were Newcastle colliers, conscripted for the voyage. In addition, England's Dutch allies sent fifteen ships commanded by William of Nassau. On paper the fleet looked impressive, but appearances were deceptive. Some of the English ships were survivors of the Armada, old, slow and with poor, rotten and ragged sails.

The fleet transported an army of ten thousand men, although the word 'army' is perhaps inappropriate to describe an undisciplined rabble of pressed men – the dregs of society, unfed, untrained, and badly clothed. Anyone more clear-sighted than Buckingham would have seen them for what they were. Complaints from junior officers of the men's 'miserable condition for want of clothes', and the fact that many had 'nothing to cover their nakedness', had no affect on him. When the local farmers of Dorset and Devon, on whom the men were billeted before their embarkation for Spain, realized that they had no money to buy their food, they refused to feed them. Riots broke out as destitute soldiers took the law into their own hands, roaming the countryside, killing sheep and threatening violence. There was no alternative but to strip them of their weapons, with the result that few of the soldiers received any weapons training at all.

The quality of the recruits was deplorable. Contemporary writers described them as 'the shadows of soldiers', 'gaol birds', 'a rabble of poor rascals' and 'the scum of our provinces'. A senior officer reported that out of 2,500 men surveyed, 200 were physically defective, 24 seriously ill, 26 aged over sixty, four men blind, one a minister of the church, one a raving madman, several simpletons and others deformed or maimed, including one with no toes to his feet and another with one leg shorter than the other by nine inches. The system of recruiting soldiers was being seriously abused. One old and blind man was pressed simply because he had given evidence against the constable's brother, who had sworn to be avenged. A man of 60, with eleven children, was chosen because of a long-standing disagreement with a town alderman. Money often changed hands so that the rich were able to buy themselves out and the weak or the helpless poor were sent in their stead.

The fleet set sail from Plymouth on 8 October 1625. Only a few miles out one of the king's ships, the *Lion*, was found to be leaking like a sieve and had to be sent home. Further problems were to follow as the incompetence of the victualling and ordnance departments became apparent. Many ships had been inadequately provisioned and, within three days of sailing, rations had to be reduced by a quarter. To add to the misery of empty stomachs the weather seemed determined to wreak havoc. Ships sank in the high winds and Cecil's flagship, the *Anne Royal*, was badly smashed about when her cannons broke loose and swung freely about the decks. Soon the flagship had flooded holds and was only kept afloat by constant pumping.

When the weather abated, Cecil called a council of war to collect damage reports. He could little have imagined what was to happen next. The captains came to him with complaints of every sort, real or imagined, and the commander of another king's ship, the *Dreadnought*, claiming to be badly holed, tried to leave the fleet to return home. Only Cecil's desperate entreaties kept her with the fleet.

Cecil was horrified to learn that not only were the gunpowder and food soaked, and the fresh water contaminated, but many of the army's armaments were defective. The muskets had been badly made and had no touch-holes to light, while the shot supplied did not fit the firearms and the moulds for making new shot were warped.

During the storms contact had been lost with the Earl of Essex's squadron of some forty ships. This was alarming in itself, but Cecil's worries were made worse when a large force of what appeared to be enemy ships were sighted as they drew near to the coast of Spain. Cecil ordered some of his fastest ships to chase the fleeing enemy in the hope that they might be part of the Spanish treasure fleet due from South America. After a tense and exciting chase it was discovered that the 'flying enemy' were in fact some of Essex's ships that had carelessly failed to signal their identity. Essex's disobedience of Cecil was to be a feature of the entire expedition.

Arriving off Cadiz, Cecil ordered Essex to enter the harbour and find an anchorage for the fleet. Ignoring his orders, Essex sailed into the harbour and single-handedly attacked twelve Spanish galleons with some fifteen galleys alongside. Only the arrival of the rest of the fleet saved Essex from destruction. On seeing the size – if not the quality – of the English fleet the Spaniards fled up a narrow creek. Cecil was keen to follow them immediately and win a great victory, but his naval advisers assured him the Spaniards could not escape from the creek. They were trapped there and the English could destroy them at their leisure.

Meanwhile, an English merchant captain had escaped from Cadiz to tell Cecil that the Spaniards had been taken unawares and that the city was weakly garrisoned. An immediate attack would take the city. But again Cecil listened to his sea captains, who told him that Cadiz must wait. They insisted that the fort of Puntal must be bombarded into submission before the ships could rest safely at anchor.

As darkness fell twenty Newcastle colliers of shallow draught joined five Dutch warships in an attack on the fort, while three English men-of-war, which drew too much water to go inshore, formed a second line of attack. At once all was confusion. While the Dutch ships bombarded the fort, the conscripted colliers, disobeying orders, remained calmly at anchor, their captains unwilling to risk the ships that were their livelihood. In the exchange of fire with the fort the Dutch suffered heavy casualties and one of their ships ran aground. The Dutch admiral was furious with the collier captains, and when the attack was renewed by Essex's squadron the colliers still hung back. When they were at last goaded into fighting they fired a few long shots, one of which went through the stern of Essex's flagship. In frustration Essex ordered them to stop firing. Perhaps they had made their point. When the fort was eventually taken after a botched landing of soldiers, it was discovered that in spite of firing over 2,000 rounds of ammunition the English warships had failed to damage the fort.

The capture of Fort Puntal took 24 hours. The bombardment had roused the entire countryside and Spanish troops were now rushing towards Cadiz from Seville, Malaga, Gibraltar and Lisbon. Oblivious to all this, Cecil ordered a general disembarkation of his troops. When news reached him that Spanish troops had been sighted at the Zuazo Bridge he immediately led 8,000 men to try to cut off the enemy.

The island of León, across which the English troops were marching, was flat and marshy, given over mainly to producing salt. As they marched the soldiers must have been puzzled by the curious conical-shaped pillars of salt collected from the watery swamps that made up much of the island. The very atmosphere seemed to be impregnated with salt, and the soldiers, toiling in the heat of the Spanish sun, built up a dreadful thirst.

When the soldiers halted for a rest, one of the regimental commanders told Cecil that his men had brought no provisions with them and had not eaten since landing. Cecil had no alternative but to send them and the men of another regiment back to Puntal. With darkness approaching Cecil ordered his men to camp for the night in an open field, near some deserted buildings. These, it transpired, were used for storing wine. At once the men began to complain that they had brought no water with them. As a humane gesture Cecil ordered that one butt of wine was to be broken open for the use of each regiment. But the salty atmosphere, the effects of the heat, the men's raging thirst and their empty stomachs produced a not very surprising result. The army rapidly degenerated into a raging, drunken mob. The half-famished soldiers broke into the wine store, smashing

open the casks. Discipline collapsed with men firing wildly into the air, shooting each other and threatening any officer who tried to restrain them. When Cecil ordered the wine to be tipped away, men scooped it up in their helmets and then besieged the commander's room, threatening his life. For his protection Cecil's bodyguard had to fire into the mob. As Cecil said later, 'The enemy with three hundred men could have routed us and cut out throats.' For the rest of that dismal night the officers had to protect themselves from the attacks of their own men.

The rising sun heralded the morning after the night before. Cecil had no alternative but to return to Puntal, leaving a hundred or so men still drunk in the ditches, to be butchered later by the Spanish. The road back to Puntal was littered with weapons and equipment, abandoned by the hungover soldiers. On his return to his flagship Cecil met with further disappointment. The Spanish ships that had escaped into the narrow creek had used the opportunity to sink four hulks across the entrance, so that the English warships would only be able to enter one at a time. Cecil fumed at the loss of such a great opportunity to weaken the Spanish navy.

Little was left for the English but to retreat in good order. Even this was denied them as his soldiers, inadequately fed and clothed, were forced to spend a further night in the fields outside Cadiz. Again the elements seemed to be mocking them as storms and sheeting rain made it a night of misery. The next day the thoroughly demoralized troops re-embarked and the fleet swept out of Cadiz Bay to the joy of Cadiz's citizens.

On the voyage home Cecil revealed an incredible lack of common sense. Plague had broken out in the crowded fleet and some crews were so badly hit that they were unable to sail their ships through the rising storms. Cecil's solution was to order thirty of his ships to send two healthy men each to the ships where the plague was heaviest and in return take two of the sick men. The effect of this, not surprisingly, was to spread the infection quickly round the fleet.

At this stage the weather worsened and the English fleet was scattered by storms. Cecil's flagship was so badly damaged that she limped into Kinsale with six feet of water in her holds and 160 sick crew in addition to 130 dead bodies. Through the winter of 1625–6 the English ships limped home to discharge their wretched human cargoes into the coastal towns of the south of England. While Cecil and Essex travelled to report to their master, Buckingham, at the king's court, the real casualties of war ended their days in squalor and misery. One commander described the condition of his men on their return to England: 'They stink as they go, the poor rags they have are rotten and ready to fall if they are touched. The soldiers are sick and naked, and the officers moneyless and friendless, not able to feed themselves.'

THE EXPEDITION TO WALCHEREN (1809)

The expedition Britain sent to the Netherlands in 1809 was one of the largest combined operations ever seen up to that time. It consisted of seventy thousand soldiers and sailors in 616 ships, 352 of which were transports and 264 warships. This prodigious military effort was intended to serve two purposes: to destroy French naval power in the Scheldt River, and to form a diversion to help the Austrians, engaged at that moment in a desperate struggle with Napoleon on the Danube. If the aims were worthy, the execution was less so. A contemporary newspaper printed a jingle which summed up popular reactions to both the military commander, Lord Chatham, and his naval equivalent, Sir Richard Strachan:

> Lord Chatham with his sword undrawn,
> Kept waiting for Sir Richard Strachan;
> Sir Richard, eager to be at 'em,
> Kept waiting too – for whom? Lord Chatham!

In fact, there was a lot of truth in this rhyme. The British commanders were an ill-assorted pair; Chatham too cautious by half and Strachan too stormy. Chatham, elder brother of the late prime minister, William Pitt, and a personal friend of the king, was a cabinet minister himself and his was clearly a political appointment. His reputation for extreme caution, which had earned him the nickname 'the late' Lord Chatham, made him a peculiar choice for an operation needing quick thinking and instant execution. It also made him the most unsuitable companion for Sir Richard Strachan, known in the navy as 'Mad Dick' because of his

wild temper and tempestuous ways. Strachan was described at the time as 'extremely brave and full of zeal and ardour . . . an irregular, impetuous fellow'.

Nevertheless, the twin aims of the expedition had much to recommend them. The Scheldt contained a great concentration of French naval power, second only to Toulon, and was described by Napoleon as 'a cocked pistol pointed at the head of England'. It was taken so seriously by the Admiralty that two separate British naval squadrons patrolled the Dutch coast at all times. Certainly a surprise attack and the burning of French ships and installations would be a major setback for French plans to invade England.

The expedition's second aim – to help the Austrians fighting Napoleon on the Danube – was more difficult to achieve. It depended on the Austrians staying in the fight long enough for the British presence in the Scheldt to have an effect on French troop allocations. When news of the Archduke Charles's victory over Napoleon at Aspern-Essling reached England it was greeted with great joy and seemed in itself to justify the Scheldt expedition. However, before the fleet could sail, news was received of Napoleon's war-winning victory at Wagram. With Austria negotiating a new peace, one of the expedition's aims was lost already. Was the destruction of the French fleet worth the risk and expense of the expedition? The decision to proceed was a hard one, but it might have been justified if Antwerp had been captured. After all, now he had defeated Austria, Napoleon would have more time to give to his proposed invasion of England.

What followed was an extraordinary saga of muddle and confusion. Chatham's failure to advance and capture Antwerp seems less important now than the fact that the authorities sent the best part of the nation's military forces to campaign in one of the unhealthiest parts of Europe without making adequate medical provision. Napoleon need not have worried. Disease was to win this campaign for him without the assistance of the French army.

The island of Walcheren, which was captured by the British in 1809, was renowned for its insalubrious climate. A few years before a French commander had lost 80 per cent of his men to the dreaded local fever. The British authorities were quite aware that in this low-lying land, reclaimed from the sea only by a system of dykes, there were many swampy areas with stagnant pools in which mosquitoes bred in their millions. The inhabitants of Walcheren were known to be 'pale and listless, suffering much from scrofula, the children rickety and all much deformed'. Heavy floods the previous year had made conditions ideal for mosquitoes and by the summer of 1809 the situation was ripe for a disaster.

British troops had encountered a similar fever in the Low Countries on a previous occasion. As early as 1747 Sir John Pringle had written about a fever that struck British troops at Antwerp.

> The sickness never begins till the heats have continued long enough to give time for the putrefaction and evaporation of the water. The epidemics of this country may therefore be generally dated from the end of July or the beginning of August . . . their decline, about the first fall of leaf; and their end, when the frost begins.

And yet the expedition went ahead in spite of the likelihood, almost certainty, that the troops would be struck by the fever. Indeed everything seemed stacked against the British soldiers once they landed on Walcheren. The weather in August 1809 was very hot and steamy, with frequent thunderstorms. Thick mist rose above the swamps and mosquitoes plagued the British soldiers as they laboured on defensive positions. Nothing, it seemed, could keep off the mosquitoes, and men became listless and careless, yawning a lot and finally collapsing. When the fever developed, the men suffered from a burning thirst and high temperatures – the classic symptoms of malaria.

While the commanders quarrelled the men died, their enemies not the French soldiers assembled on the other side of the Scheldt River, but the insignificant, unconsidered insects. (It was believed at that time that malaria was caused by mist or bad air – hence the name of the sickness.) Between 6 August and 3 September, the number of cases increased from 688 to 8,134. The antiquated medical services of the British army were quite incapable of coping. Doctors fought the epidemic with quinine, but their supplies of tree bark were hopelessly inadequate. In many cases they too succumbed.

The Medical Board back in England had not even been consulted when the campaign was planned, and the number of doctors sent on so large an expedition was ridiculously low. There were no facilities to cope with so many casualties at one time and men were forced to lie on the beaches in their own filth. The men suffering from fever needed to be returned to England, away from the insalubrious climate of Walcheren, yet the position was little better once they got back home. Hospitals were simply deluged with cases of Scheldt fever. By October, 1809, only 5,616 men on Walcheren were fit for duty, out of an original force of 40,000 soldiers. Just 106 men had died in combat, while over 4,000 died from fever. The French had to do nothing but wait; they had never had so easy a victory.

The Walcheren expedition was a tragedy. The force sent under Chatham and Strachan contained a galaxy of junior officers who were later to earn fame with Wellington in Spain. Yet they had no chance to show their skills. Even with a massive naval force, the British achieved only limited success in harming French shipping and installations. Although Napoleon later admitted to suffering losses of £2,000,000, the expedition had cost five times that much. And the losses in men were irreplaceable; 4,000 dead and 12,000 so badly affected by the fever that they would never serve again.

THE MY LAI MASSACRE (1968)

When the case of Lieutenant William L. Calley, who ordered the massacre of hundreds of Vietnamese villagers at My Lai in 1968, became public knowledge, America broke out in a rash of self-disgust. Suddenly the phrase 'war crimes', once considered only to apply to the actions of the likes of Adolf Eichmann and Heinrich Himmler, was on everyone's lips. Did Americans really do things like that? The answer was an emphatic 'yes'. One World War II general even wrote, 'If Germany had won, I would have been on trial at Nuremberg instead of the Krauts.' The press began digging for dirt in America's past – and found it. Calley was not unique. What about 'Howling Jake' Smith?

On 28 September 1901, in the town of Balangiga in the Philippines, 74 American servicemen were taking breakfast. They were surprised by a band of Filipino guerrillas who mowed down the unarmed soldiers as they waited in a mess queue. Thirty-eight Americans died and a further eleven were wounded, some of them mutilated by their attackers. The news shocked Americans, most of whom had believed the nationalist rising in the Philippines was over. In the highly charged atmosphere, General Jacob Smith was given the task of avenging the massacre and bringing the culprits to justice. Unfortunately, Smith allowed his emotions to get the better of his judgement, issuing the famous order, 'Kill and burn and the more you kill and burn the better you will please me.' Smith said that he wanted no prisoners; anyone over the age of ten was old enough to be a rebel and to suffer the consequences. He intended to turn Samar Island – the centre of Filipino resistance – into a 'howling wilderness'. Smith's savagery predated the Nazis by 40 years. When an American soldier was killed, Smith selected a Filipino by lot and he was executed in retaliation. By such crude but effective methods Smith succeeded in quelling the Filipino rising.

During a march through the central part of Samar, Major Littleton Waller ordered the execution of eleven native guides on the grounds that they had found some edible roots which they had concealed from the starving Americans. Waller was court-martialled, but at his trial he defended himself by claiming that he was only following General Smith's instruction to kill and take no prisoners. Waller was exonerated, but when the American press became aware of the case and 'Howling Jake's' order they demanded the general's court martial. The *New York Evening Post* even advocated his execution if found guilty of atrocities. On a wave of popular resentment, 'Howling Jake' was court-martialled and dismissed.

For nearly seventy years the story of Jake Smith remained like a skeleton in the cupboard, always threatening to emerge to challenge the American people's perception of their armed forces as squeaky clean crusaders in a world of filth and barbarism. But it was in Vietnam that the bubble really burst and it was Lieutenant William Calley who provided the pin.

During the late sixties morale among American soldiers in Vietnam was so bad that the authorities needed to maintain a fast rotation of troops and officers to ensure that no one stayed there too long. As a result, commissions were granted to some men who could never have hoped to have commanded troops in any other conflict in American history. One such man was Second Lieutenant William Calley, Jr, a 'below-average, dull and inconspicuous boy', who could not read a map properly even after he had graduated from Officer Candidate School. Only in unusual circumstances – and the Vietnam War was one of them – would a man of Calley's low intelligence have been given command over men in battle.

In the early months of 1968 'Charlie' company of 1st Battalion, 20th Infantry, to which Calley was attached, had suffered heavy casualties and felt that they had a score to settle with the Vietcong (South Vietnamese Communists). On 16 March Calley's platoon was ordered to enter the village of My Lai in search of Vietcong. They found none, but Calley's company now indulged in a sickening orgy of cold-blooded killing. Old men were stabbed with bayonets, women and young girls raped and then shot in the back of the head or blown to pieces by grenades. A total of 347 Vietnamese villagers were massacred. Some of Calley's soldiers were unwilling to carry out his orders at first, but the pressure to conform was too great. Like Nazi concentration camp commandants in the 1940s, they later claimed that they were just following orders. One soldier described his experiences of killing men, women, children and even babies:

And we huddled them up. We made them squat down and Lieutenant Calley came over and said, 'You know what to do with them, don't you?' And I said yes. So I took it for granted that he just wanted us to watch them. And he left, and came back about ten or fifteen minutes later and said, 'How come you ain't killed them yet?' And I told him that I didn't think you wanted us to kill them, that you just wanted us to guard them. He said, 'No, I want them dead.'

So I started shooting. I poured about four clips into the group . . .

Yet not all Americans had lost their sense of morality that day. At one point Warrant Officer Hugh Thompson, observing from a helicopter, landed in My Lai and threatened to open fire on Calley's men unless they ceased the killing.

For a year the Army authorities maintained a cover-up. Only when a news reporter released photographs showing details of the dreadful massacre was an enquiry set in motion. Calley was later court-martialled, like Smith 70 years before, and found guilty of murdering at least twenty-two South Vietnamese civilians. Sentenced to hard labour for life, Calley received more lenient treatment than he deserved when President Nixon intervened to reduce his sentence to three years' house arrest at Fort Benning, with the privilege of having his girlfriend visit him regularly.

Yet for many Americans Calley became a hero. The governor of Indiana ordered all flags on state buildings to fly at half mast because of Calley's conviction. At Fort Benning the Reverend Michael Lord told a church congregation, 'There was a crucifixion 2,000 years ago of a man named Jesus Christ, I don't think we need another crucifixion of a man named Rusty Calley.' A record entitled 'The Battle Hymn of Lt. William Calley', in which Calley met the great commander in the sky and told him that he had followed orders and done his duty, enjoyed much success.

Lieutenant Calley was just one of many war criminals in Vietnam and not all of them were soldiers. Was he just the thin end of a very thick wedge? How high in the military and political administrations did knowledge of the cover-up go? From the way in which President Nixon reacted, we can assume it went right to the top. For America, My Lai was a military disaster every bit as real as Pearl Harbor or the Kasserine Pass. My Lai and atrocities like it eventually convinced the American people that the war in Vietnam was not worth winning, that the American cause was not the right one and that people should perhaps have the right to choose between being 'red or dead'.

Clearly the training of American soldiers in Vietnam had been seriously at fault. They had not been prepared for the enormous culture shock involved in leaving the affluence of modern American urban society for the poverty of Southeast Asia. Inherent racialism caused many 'good American boys' to regard the 'gooks' (Vietnamese) as inferior, and behave in ways which Middle America could not believe possible. The mother of a man from Calley's platoon summed up the problem when she said, 'I sent them a good boy, and they made him a murderer.' Calley's own comment on the massacre at My Lai – '. . . no big deal, Sir' – deserves a prominent place as an illustration of the dehumanizing effect of warfare.

CHAPTER 7: PLANNING FOR DISASTER

The Luddite Tendency

The pace of technological change in the twentieth century has presented military planners with a range of complex problems. Their reaction – or lack of it – when new weapons became available has contributed to the military success or failure of many commanders. During World War I two weapons – the machine gun and the tank – provoked a strong reaction from traditionalists who believed in 'cold steel' and the horse rather than machines.

Two per battalion

In 1914 the machine gun was hardly a new weapon. It had been in use in most European armies for more than forty years. Yet its impact had been small and it was felt to be most suitable for colonial wars, where it was remarkably cost-effective in terms of the small number of white troops needed to kill the maximum number of rebellious natives. This equation, it was felt, would not apply in a war between the white races. Yet although a

The 9th Lancers charge a German battery at Mons in 1914. Many British generals – particularly cavalrymen like French and Haig – believed that cavalry would play a significant role in the fighting in World War I. But with the advent of trench warfare the cavalry became an expensive liability.

report in 1871 described it as a useful weapon for the British, with their relatively small army, little was done to equip the British Expeditionary Force (BEF) with an adequate number in 1914. Many arguments were used against the guns: they were too expensive, used too much ammunition, were too heavy and, most damning of all, they encouraged defensive thinking in an age which saw the offensive as the only acceptable tactics. Many soldiers felt the machine gun would literally slow down the infantry advances and reduce the mobility of the army as a whole. The British would appear to have learned nothing from the Second Boer War. During that conflict the defensive firepower of the Boers inflicted defeat after defeat on Buller with his inflexible 'up and at 'em' approach (see p. 26), yet was overlooked as untypical of what could be expected in a European war. Instead military thinkers cited the wateful and barbarous 'human wave' methods employed by Japanese general Nogi in the Russo-Japanese War as proof of the efficacy of offensive tactics (see p. 112). The idea that Nogi was perhaps butchering his men unnecessarily does not seem to have occurred to them. In 1914 General Altham reached the staggering conclusion that 'the Manchurian campaign has wiped out the mistaken inference from South African experiences that bayonet fighting belonged to the past'.

'The machine gun is a much overrated weapon, two per battalion is more than sufficient.'

General Douglas Haig, 1915

Haig's famous gaffe, one of the worst miscalculations of the war, was a product of sheer bloody-mindedness. Still obsessed with the war of movement, Haig was unwilling to equip himself properly for the static defence which was imposed on both sides by the terrain and the state of military technology. Haig saw the machine gun as a defensive weapon which would slow him down when the time came to attack.

Sir Douglas Haig played a significant role in delaying the introduction and limiting the mass use of the machine gun. Just as he condemned hundreds of thousands of his own men to death at the hands of German machine gunners, so many Germans owe their lives to the fact that he limited the number used by British battalions in the first two years of the war. In 1909 Haig insisted that the machine gun was a defensive weapon, encouraged defensive thinking, and made men passive and defeatist in outlook. The persuasive arguments of the noted military theorist, Major General Fuller, in favour of an offensive based on superior firepower and penetration in which the machine gun would play an active part were totally ignored.

At Mons in 1914 the concentrated firepower of the BEF so impressed the Germans that they believed they were facing massed machine guns. In fact the BEF had no more than two machine guns per battalion, and were principally reliant on rifle fire. More machine guns would clearly have enhanced yet further the effectiveness of the BEF, but French and Haig would not release men to train in the use of the few machine guns they did have. In February 1915, 890 of them were lying idle in France for lack of men who knew how to use them. Only when David Lloyd George became Minister of Munitions did matters change, the Welsh Wizard increasing the number of guns from two to sixteen per battalion. However, the introduction of the weapons was bedevilled by production problems.

The massive increase in demand took Vickers, the normal War Office supplier and hitherto acccustomed to dealing with a more modest level of machine gun production, completely unawares. Between 1904 and 1914 Vickers had been asked to build only eleven guns a year by the British and could only justify staying in production because of Russian orders for the gun. In August 1914, 192 guns were ordered, more than Vickers' capacity to produce them, followed by 100 more in September and nearly 1,500 in October. Not surprisingly the company failed to produce the full order. It is baffling that no one appears to have considered the number of machine guns that might be required in the event of a European war and ensured that there were sources of production available. Comparison with levels of production in France and Germany could have provided a yardstick, at least at the start. But as so often in British military history, conservatives delayed the introduction of vital, war-winning equipment and it was the soldiers who suffered.

The land ironclad

The story of the tank is no happier than that of the machine gun. The grim attritional struggles on the Western Front between 1915 and 1917 had produced casualty lists unparalleled in history. Mobility had only returned to warfare after the effective introduction of the tank on the ground and its combined use with aircraft. The tank had not been the 'war-winning' weapon that had been hoped, but it had shown the way forward. Future developments in tank design would surely overcome the problems it had faced in

Tanks – such as this one shown in action at Cambrai in 1917 – helped restore mobility to warfare. But conservative military thinkers in Britain failed to see their huge potential.

1917–18. Yet the men who dominated British military thinking were of Boer War vintage and still thought in terms of cavalry. Their confidence in the horse had not been dented by the fighting in Flanders. They dismissed the four-year trench deadlock of the Western Front as untypical, and drew examples from the successful use of cavalry in Allenby's campaigns against the Turks in Palestine. This extraordinary claim was, in effect, an attempt to put the clock back by half a century. Cavalry had been obsolete in European warfare since the Franco-Prussian war. There was no place for the horseman in a main battle line dominated by machine guns and rifles which could fire fifteen shots a minute, at ranges of up to a mile. The cavalryman was just too big a target. Moreover, cavalry were not cost effective as had been shown by the experience of the Russians in 1914. Forty trains were needed to transport a cavalry division of 4,000 men and twelve guns, whereas the same number of trains could move 16,000 infantry with 54 guns. With every horse needing twelve pounds of grain a day, much of the transport system was tied up in bringing forward fodder for the horses.

'The tank was a freak. The circumstances which called it into existence were exceptional and are not likely to recur. If they do, they can be dealt with by other means.'

Major-General Sir Louis Jackson, 1919

This short-sighted view was typical of British military thinking after World War I and resulted in Britain falling far behind France, Russia and Germany in tank development.

On the Western Front, the role of the British cavalry had been to exploit breakthroughs by the infantry. Yet the breakthroughs never came. And even if they had, the muddy or swampy terrain would have prevented the cavalry from breaking out into open country. But such problems did not prevent the cavalry generals from envisaging a similar follow-up role for themselves in the aftermath of a breakthrough by tanks rather than the footslogging infantry. The idea was tried at Cambrai in 1917. The tanks achieved a breakthrough, but the cavalry – lacking armour plating! – failed to negotiate the mud, barbed wire and machine guns. The same flawed tactic was implemented at Amiens on 8 August 1918. Here 600 British tanks achieved a complete breakthrough, but again the horsemen achieved little. In hindsight the whole cavalry corps would have been better employed fighting on foot, serving the guns or acting as stretcher carriers. Their meagre achievements did not justify the considerable national expenditure on them.

However, the cavalry officer did not end up as an interesting exhibit of an extinct species in the Imperial War Museum. Instead he and his kind continued to dominate British military thinking after the war. Led by Earl Haig, they did whatever they could to hinder the development of the tank and make certain that cuts in military spending did not affect the cavalry. Thus in the 1922 defence cuts Britain's army was reduced to 126 infantry battalions, 20 cavalry regiments and just six tank battalions. No sooner had peace come than tank production actually ceased. By 1929 expenditure on petrol for tanks and motorized vehicles was £72,000, that on fodder for horses £607,000.

Ironically, the most able proponents of tank warfare were British. Captain Liddell Hart and Major General Fuller may claim with some justification to have originated the

Field Marshal Sir Douglas Haig was every inch the cavalryman, but it is said of him – somewhat cruelly – that he was bright only to the top of his boots. His obsession with the role of the horse in warfare had a malign influence on the development of modern mobile warfare in Britain.

'blitzkrieg' tactics later developed and successfully used by the German general Guderian. Yet in Britain all the two men met was opposition and ostracism. The Chief of the Imperial General Staff, General Montgomery-Massingberd, actually denounced Fuller's books even though he admitted he had never read them. General Edmonds told Liddell Hart that the days of the tank were past and that if one of the tin monsters showed its face on a modern battlefield it would immediately be knocked out. Even as late as 1936, with both France and Germany developing élite tank forces, the prevailing attitude in Britain was that the role of the tank was of secondary importance.

When war came in 1939 Britain was leagues behind the Germans in tank development and strategy. The BEF that was sent to France – and described by one general as being as well equipped as possible – took just one under-strength tank brigade, one and a half battalions of which had just the Mark I infantry tank, equipped with a machine gun only. Only half a battalion was equipped with Matilda tanks and even then no real match for the best German, Russian and French tanks – a shameful state of affairs for the nation which had invented the tank and, through Fuller and Liddell Hart, had developed the philosophy of modern tank warfare. Yet her army went to war with derisory tank strength and was overwhelmed by the armoured forces of Germans like Guderian, who had learned the right lessons from defeat, just as the British had learned the wrong lessons from victory.

Adapt or Die

Intelligence failures lie at the root of many military disasters. Yet sometimes the problem is not the lack of information but the fact that the crucial data only arrives after the planners have already decided on their plan of action. At this point they must decide whether to change their minds in the light of the new data or else go ahead regardless. Inability or refusal to adapt to a changed situation was the cause of three of the most significant military blunders of World War II.

The impassable Ardennes

French strategy in 1940 was based on the mistaken belief that the Germans would attempt a re-run of the Schlieffen Plan of 1914, moving through Belgium to invade France from the northeast. The French were certain that their massive defences in the Maginot Line would protect their eastern frontier with Germany and that therefore there was no other way for the Germans to come. French planners believed that if the cream of the French Army advanced into Belgium to meet the Germans on the 'Dyle Line', France itself would not be invaded. So strong was the French belief in this scenario that at the 'hinge' of the French line, along a hundred miles of the thickly wooded hills of the Ardennes, which they believed to be impenetrable, there were just fourteen weak divisions. On the other side of the forest, unknown to the French, were 45 German infantry divisions and 10 panzer armoured divisions, under Field Marshal von Rundstedt.

'The Ardennes are impenetrable . . . this sector is not dangerous.'

Field Marshal Henri Pétain, 1934

This fatal judgement – flying in the face of historical evidence – was to render the Maginot Line and Gamelin's Dyle Plan of 1940 completely pointless. When the Germans attacked through the Ardennes they found the area denuded of French divisions.

History shows that far from being impenetrable, the Ardennes had provided a way in to France for invaders no fewer than ten times in the 200 years prior to World War II. For the French to ignore this is baffling enough; more incredible still is the revelation that during French army manoeuvres in 1938, General Prétalat used exactly the plan the Germans were to follow in May 1940 and succeeded in breaking through the French defences with just seven infantry and two armoured divisions. What Rundstedt would do with his massive force would soon be seen. Instead of learning from Prétalat's achievement, Gamelin, the French commander-in-chief, decided to suppress reports of the fiasco, insisting that in the context of a real invasion reserves would be available to stop such a breakthrough.

Gamelin was insistent in the view that the main German strike would come through Belgium and closed his mind to any evidence to the contrary. And there was a growing fund of evidence to show that he was wrong. In contrast with Gamelin's interpretation of German intentions, French Military Intelligence (the Deuxième Bureau) had for several months been noting the German military build-up in the area around Sedan. Gamelin called this build-up a feint, intended to mislead them into thinking the Germans would strike in the Ardennes. But intelligence sources had by now located all the German armoured divisions and they were undoubtedly not heading for Belgium. Swiss sources reported that the Germans had built eight new bridges across the Rhine, all indicating that the Ardennes were the target. The French military attaché in Berne even got the date right, reporting an attack towards Sedan between 8 and 10 May. And yet Gamelin refused to listen. He was so determined to avoid the horrors of 1914–18 and the devastation of northern France that he was prepared to run the risk of being taken at a disadvantage by the Germans and having his army cut in two. The débâcle of May 1940 was avoidable; the French High Command blundered to defeat because it was not prepared to alter its preconceptions in the light of new intelligence.

'There is no such thing as the aerial battle. There is only the battle on the ground.'

General Maurice Gamelin, 1936

Even as late as 1936 the French – and British for that matter – did not appreciate the value of the strategical use of airpower to support ground troops, the essence of German blitzkrieg tactics.

The camera never lies

Operation Market Garden, the airborne assault on the Rhine bridges in 1944, was a brilliant operation which failed by trying to go 'a bridge too far'. Nevertheless, human error played a prominent part in the failures in the Arnhem sector. The planning of the operation involved the collection of an enormous amount of aerial photo reconnaissance in order to ascertain what sort of opposition the airborne troops might meet once they had landed. Field Marshal Montgomery was convinced that any opposition would be second rate and when he was warned that there might be panzer divisions in the area of Arnhem he ridiculed the idea, saying his men had more to fear from the terrain than the Germans. Yet intelligence reports from Dutch sources claimed the 9th and 10th Panzer divisions were near Arnhem. How serious a threat might these divisions prove? The planners decided that they were being re-equipped and that few of their tanks were operational. But when photographic evidence was brought in it clearly showed operational German tanks in the area – a revelation which called the wisdom of the whole venture into question. However, Lieutenant-General Browning, head of the British First Airborne Corps, elected to ignore the evidence of the photographs. His famous

The bridge at Arnhem, where British airborne troops fought a heroic but doomed action against overwhelming German armoured forces in 1944. British aerial reconnaissance had warned of the presence of German armour in the area, but their reports were ignored.

words to the intelligence officer who showed him the pictures, 'I wouldn't trouble yourself about those if I were you', have echoed down the years. When the young officer tried to insist that the matter be taken more seriously he was suspended from duties by the corps medical officer. Having effectively silenced the opposition, Browning went ahead with the drop at Arnhem, with disastrous consequences.

German strength around Arnhem was, in reality, considerable. The XI SS Panzer Corps of General Bittrich, including 9th and 10th Panzer divisions as reported by the Dutch, were there. Moreover, both Field Marshal Model and General Student were in the area, and took immediate counter-measures against the British landings, partly as a result of finding a full set of British plans on the body of an officer killed in a crashed glider. The advance of the British paratroopers into Arnhem was held up by strong German resistance and only the 2nd Parachute battalion reached the Arnhem bridge, where they fought a heroic but hopeless battle against overwhelming German forces. The inadequacy of British radios led to a total breakdown in communications between the isolated parts of the British division and poor weather prevented support troops being flown in until it was too late. Eventually, Montgomery ordered an evacuation of British and Polish troops from Arnhem, claiming the operation was a 90 per cent success, a curious interpretation of a heavy and avoidable defeat. Of 10,000 airborne troops dropped around Arnhem, only 2,163 were rescued. Of the others, 1,130 were killed and over 6,500 captured, half in a wounded condition. The Germans reported losses of some 3,300.

A blip on the screen

The Japanese attack on Pearl Harbor on 7 December 1941 achieved such total surprise that it has been lengthily debated since as to whether President Roosevelt allowed the

Japanese to strike first to allow him to bring a united USA into a war against Japan and Germany. Whatever the truth – and it may never be known – the American commanders in Hawaii knew nothing of what was about to hit them. In spite of growing tension in the Pacific between the USA and Japan, the authorities at Pearl Harbor had taken only half-measures to protect themselves against that well-known Japanese speciality – the surprise attack.

The Japanese struck on a Sunday, possibly on the assumption that for Anglo-Saxons it is a day of rest and commanders would therefore not be at their posts. With 96 American warships in harbour, and with the threat of war very real, one might have supposed that the Japanese would be wrong in their assumptions. They were not. Admiral Kimmel, commander-in-chief of the American Pacific Fleet, was playing golf with General Short. Servicemen were off duty and an air of relaxation was apparent everywhere. Intelligence reports had already indicated substantial Japanese naval activity, but the Americans had complacently decided it was headed for somewhere in Southeast Asia. Kimmel himself was convinced that the Japanese would not attack Honolulu. When an American radar station on the northern tip of one of the Hawaiian islands, Oahu, reported unexplained blips on the radar screen, the duty officer, Lieutenant Kermit Tyler, dismissed the idea that they might be Japanese bombers with the words, 'Well, don't worry about it.' Tyler was expecting a flight of B-17 bombers and could not believe that the blips on the screen might be hostile planes. A call to action at that stage would not have saved Pearl Harbor completely – it was already too late for that – but it would have saved hundreds of lives. Thirty minutes later Japanese planes were bombing

Burning oil from the shattered fuel tanks of a damaged American battleship turns Pearl Harbor into a sheet of flame after the surprise Japanese attack on 7 December 1941.

American ships and killing American servicemen. In those thirty minutes crews could have gone to action stations, guns could have been manned, fighters could have taken off and met the incoming Japanese bombers. America paid a high price for Tyler's complacency. But the real responsibility lies with politicians in Washington and with senior commanders in Honolulu who were blind to intelligence reports.

'We are not Poles. It could not happen here.'

General Maurice Gamelin 1940, on receiving a warning that the Germans might use blitzkrieg tactics against France as they had in Poland

It did happen there and France suffered a collapse even more humiliating than that of Poland.

AUSTRIAN MOBILIZATION (1914)

The outbreak of war in 1914 was welcomed by the military leaders of each of the great powers except one. In contrast with the supreme confidence of the French, Russian, German and British general staffs, who expected to achieve victory by Christmas, the Austrian military leaders were plunged into gloom. Even the Austrian commander-in-chief, Conrad von Hötzendorf, spoke of the war hammering in 'the coffin nails of our monarchy', and regarded the struggle against Russia as hopeless from the start. For many Austrians the Russian war machine was 'a black steamroller which will obliterate us'.

One might have expected that such negative thinking would have either persuaded Austria to avoid war altogether, or else to make certain that if it came her army was ready for it. However, Austria did neither. In the first place, she pressed for war against Serbia, even in the knowledge that it would draw Russia in against her. In the second place, her military leaders did little to equip their troops for a modern war. With sublime indifference the Austrian generals disregarded the lessons of the Russo-Japanese and Balkan wars and continued to think in nineteenth-century terms, with glittering parades of cavalry and tactics little changed from the days of Napoleon. So diverse

were the nationalities in the Austrian army that in some regiments the only common language was English, which the men had learned in the hope of emigrating to the USA.

Because the Austrian command structure, the AOK, was situated far from the front at Teschen in Austrian Silesia, Conrad and his planners were totally out of contact with their troops. At Teschen they lived a life of luxury, surrounded by the trappings of privilege – uniformed servants and candlelit dinners. Between 1914 and 1917 Conrad made just three visits to the front.

Although a Russian invasion of Austrian territory was imminent in 1914 Conrad persisted in giving priority to the invasion of Serbia. Blinded by a personal hatred of the Serbs, he relied on the Germans to hold back the Russians, even when told that Germany would be concentrating on ensuring the success of the Schlieffen Plan in the west. Austria would need all her strength to hold back the Russians on her own frontier, yet Conrad ordered the twelve divisions of the 2nd Army to travel south to face the Serbs, leaving a gaping hole in the Austrian line against Russia. Conrad seems to have felt that recalling these troops might cast doubt on the efficiency of the AOK, so he decided on a compromise: they would go to the Balkans for ten days, and then return to the Russian front. This arrangement, ridiculous though it was, might have had a chance of working if the Austrian commander on the Serbian front, General Potiorek, had been told that the 2nd Army had only been lent to him for ten days. In the event, Potiorek included the twelve divisions in his plans for the invasion of Serbia, only to have them withdrawn at the last moment, throwing his planning into confusion.

To shuffle troops as Conrad did requires a highly efficient railway system. The Germans had such a system and were thus able to rush enormous numbers of troops to the Belgian border in 1914. The response of Austria's railway planners to the army's transport needs was the stuff of bureaucratic nightmare. First they overestimated the need for troop trains and closed down virtually all commercial life throughout the country. Then, on the grounds that war imposes stricter conditions, they limited military trains to 50 carriages per locomotive, rather than the 100 carriages permitted in peacetime. In order to 'preserve a uniform pattern' in mobilization, all trains were ordered to travel only at the maximum speed of the slowest train on the worst line. As a result the average speed of Austrian troop trains was less than that of a bicycle.

Feeding arrangements for troops in transit also reflected convoluted bureaucratic thinking. Trains were halted every six hours for feeding-pauses, even though the regiments had their own field-kitchens with them on the trains. As stations possessing facilities for feeding so many men were few and far between, troops sometimes travelled all day without food only to be given two square meals in succession in the middle of the night. The pressure obviously affected some men severely. The station master at Podborze in Silesia broke down completely, reversed all his signals, held up eight troop trains for several hours, and then shot himself.

Hearing from von Moltke, the German commander-in-chief, that there was no question of a German offensive against Russia, and without even waiting for the return of the twelve divisions of the 2nd Army, Conrad launched an offensive against the Russians in Galicia. Heavily outnumbered and fighting with poor direction and outdated tactics, the Austrians lost nearly a third of their combat strength in the first three weeks of the war. This was planning for disaster in a very real sense. Only incompetence on a similar scale by the Russians prevented the defeat of the Austro-Hungarian Empire 'before the leaves fell'.

THE SUVLA BAY LANDINGS (1915)

Sir John French may have regarded the Gallipoli expedition in 1915 as a 'sideshow', but there was no need to send clowns to command it. Just as the Admiralty refused to release more than one of their front-line battleships for use in the Dardanelles, so French refused to allow his best commanders to be moved from France, where he felt the war would be decided. It is not surprising that in 'scraping the barrel' War Minister Lord Kitchener came up with 'the most abject collection of generals ever collected in one spot' to go with the antiquated pre-Dreadnoughts that comprised the naval force.

By August 1915 the original naval attack and the subsequent landings at Cape Hellas had failed to achieve a breakthrough. Six months before the emphasis had been on swift movement and decisive results, but now the campaign was as static as that on the Western Front, with complicated trench systems and a mentality which measured progress in yards. When a new landing at Suvla Bay was planned, it was hoped that in conjunction with a breakout by Anzac troops from the cove in which they had been trapped since April, this might bring the campaign to a speedy conclusion. The operation needed a most able commander, who could keep his troops to a tight timetable, but where could he be found? Not in France, certainly, for French had made that quite clear. For so senior a post there were just two candidates in Britain: Lieutenant-General Ewart and Lieutenant-General Stopford. By a narrow margin, Stopford won the command.

Stopford was just about the worst choice possible. Although only 61, he looked and behaved as if he were older (he had, in fact, been on the retired list since 1909). He had served as Buller's (see pp. 26 and 37) military secretary during the Second Boer War, but had never actually commanded any troops at all. Stopford was a knowledgeable military historian, but was unfit for active service and far too weak for the arduous conditions he would face at Gallipoli. The British commander Sir Ian Hamilton, had tried to impress on Kitchener the importance of selecting a man of strong constitution and steady nerves for the new commander of IX Corps. He is hardly to blame if Kitchener came up with the feeble Stopford.

The officers who came with Stopford were elderly men who had been passed over for any other active service through poor health or other unsuitability. Major-General Hammersley, for example, had only recently recovered from a nervous breakdown. After landing, and encountering gunfire for the first time, he lay down on the floor of his tent with his hands over his head. Not surprisingly his division failed to occupy the hills designated as their targets. Hamilton was understandably furious at Kitchener. 'I know he is not capable of understanding how he has cut his own throat, the men's throats and mine, by not sending young and up-to-date generals to run the divisions.'

The Suvla plan in its simplest form envisaged the landing of 22,000 men on a number of beaches against minimal Turkish opposition and a prompt march four miles inland to occupy a semicircle of hills, including Chocolate Hill, Scimitar Hill and Ismail Oglu Tepe. From here IX Corps would join the Anzacs in an attack on Sari Bair. Surprise and speed were essential. If successful, the plan would result in British control of the Narrows, allowing warships to pass into the Sea of Marmara. A swift end to the campaign could then be expected to follow. But it was crucial for Stopford to get his men inland and take possession of the undefended hills.

Stopford did not seem to understand the urgency of the operation and thought getting his men ashore a great achievement in itself. He told Hamilton that they must rest for a day after landing, and his chief of staff even spoke of calling up heavy artillery to bombard the Turkish positions before any advance was possible. This should not have been necessary. There were no sizeable Turkish positions to speak of, just 1,500 Turkish militiamen under a German cavalry officer, Major Willmer, standing between Stopford and one of the decisive victories of the entire war. To Willmer, watching the British landing, it was so chaotic it resembled 'a disturbed ant heap'.

Of Stopford's fellow-officers, Hammersley felt too unwell to advance, and one of the brigade commanders, named Sitwell, claimed he was too exhausted to move on to assault the hills and that his men needed a rest. Even when one of his officers reported that there were no Turks at all in his sector, Sitwell still did not occupy the hills. To

The Allied forces at Gallipoli were often ill-equipped. Their Turkish opponents were well supplied with German hand grenades, but the British War Office had not seen fit to supply grenades to their own troops. Here Australian troops improvise with home-made grenades made from tin cans.

compound these problems, one of the divisional commanders had a tantrum. Lieutenant-General Sir Brian Mahon felt himself too senior to be commanding a mere division, and when nine of his twelve brigades were allocated elsewhere, he resigned out of pique, declining to attack the 700 Turks in the hills with his 3,000 remaining men.

While Stopford was wasting his chance at Suvla Bay by inactivity, Hamilton was on the island of Imbros waiting to hear from him. He had already had reports from British flyers that there were no strong Turkish forces in the Suvla area and that therefore there was nothing to prevent IX Corps taking the hills. And yet no message came from Stopford. Eventually Hamilton sent Colonel Aspinall-Oglander and Colonel Hankey to Suvla Bay. The two men had to travel by trawler and arrived after a delay of six hours. What they found when they reached Suvla Bay was described by Hankey as 'August Bank Holiday in England', with the bay peaceful and fringed with bathers.

On board Stopford's flagship the old gentleman was 'in high spirits' and resting his leg, for he had not felt well enough to go ashore yet. When Aspinall-Oglander observed that his men had not reached the hills even though they had been ashore over 24 hours, Stopford blandly replied that they had done splendidly to get ashore safely. The two colonels had the greatest difficulty in not exploding on the spot. The chance to make a crucial breakthrough was being frittered away by a foolish old man apparently unaware that this was one of the decisive moments of the entire war. When they tried to press him about an advance, Stopford told them that he would think about an advance the next day. At this moment a message was passed to Stopford that Turkish forces were moving towards Suvla at high speed. Undeterred, he contacted his divisional commanders to see if they felt up to advancing today. His exact words were: 'In view of the want of adequate artillery support I do not want you to attack an entrenched position held in strength.' Hammersley and Mahon were not the men to miss a chance to stay put rather than go forward, and decided on a policy of masterly inactivity.

On Imbros, Hamilton, meanwhile, could feel that something was seriously wrong. He decided to follow Aspinall-Oglander and Hankey, and asked the navy for a vessel to take him to Suvla. After a five-hour delay a yacht was found; it was the best the navy could manage. In Suvla Hamilton heard the same old story from Stopford: Turkish resistance was too heavy (1,500 versus 22,000), his force had not sufficient artillery, and his men were exhausted. Hamilton, exasperated, told Stopford that his men must take the hills immediately. Stopford replied that his leg was too bad to go ashore and suggested that Hamilton talk to the brigade commanders, who might be prepared to attack. Hamilton went ashore to meet Hammersley, and ordered him to send his men forward. Baffling to recount, the British troops who had occupied Scimitar Hill on their own initiative were now recalled so that the brigades could form up on the beach and then advance a short way inland and begin to dig trenches. Scimitar Hill was simply abandoned and soon occupied by the Turks.

The next day Hamilton found that Stopford had come ashore at last but only to give instructions for the erection of a bomb-proof shelter for himself and his staff. Nearby, 800 Turks without machine guns were holding off 6,000 British troops. In one area a young officer told Hamilton, 'We are being held up by three men. There is one little man with a white beard, one man in a blue coat and one boy in shirt sleeves.'

The British had waited too long. The hills were now overrun by Turks under that formidable commander Mustapha Kemal. The assault on Ismail Oglu Tepe, which had been unoccupied throughout Stopford's leisurely sojourn in Suvla Bay, was to cost the British 8,000 casualties. After just nine days in command of IX Corps Stopford was sacked, but by now the chance of making the Gallipoli campaign a success was lost.

THE BATTLE OF GUADALAJARA
(1937)

Italian intervention in the Spanish Civil War on Franco's side was not without its problems for the Nationalists. Whereas Franco had expected Fascist volunteers to join his own regiments, Mussolini instead sent regular army and blackshirt units which fought under their own Italian commanders. It soon became clear that Mussolini's aim was less to help Franco than to win 'two or three stunning victories' which would boost his own prestige at home. As a result the co-operation between Franco's troops and the Italians was not always very good and the Spanish took a secret delight in seeing the Italians get themselves into trouble.

At the battle of Guadalajara, in September 1937, the Italian commander, General Roatta, fielded 35,000 men in three blackshirt divisions (God Willing, Black Flames and Black Arrows), as well as a division of Italian regulars. He had, in addition, 80 tanks, 200 guns and 60 aircraft. Although these forces looked impressive on paper, the regular division was partly made up of conscripted men and groups of labourers. The latter were under the impression that they were going to Africa to be used as extras in a film currently being made there, *Scipio in Africa*. The Italian troops had been equipped with thin tropical uniforms and in the unseasonable rain and sleet their morale plummeted. As fate would have it, the Italian Fascist troops were soon going to face, on the Republican side, the 12th International Brigade, spearheaded by the Garibaldi battalion of Italian anti-Fascists. Italian would be fighting Italian.

Roatta's preparations for the battle belied his status as a professional soldier. Equipped with nothing better than a 1:400,000 Michelin Road Map of the area, the commander and the staff tried to make their troop dispersals without real knowledge of the lie of the land. As a rule battalion commanders in the pro-Fascist Italian divisions had no maps at all. Logistics were neglected, with tanks immobilized through lack of fuel. It soon became apparent that many of the troops were too old and too poorly trained for the hardships of campaigning. Indeed many of the blackshirts had been turned down by the regular army for these very reasons. They lacked training in modern weapons and their officers had little experience of command. It seems that most of the Italian volunteers had no real commitment to the Nationalist cause and were only there for the money. The 175 lire they earned per week was a good wage in most of Italy, better than a labouring wage in the north and a princely sum compared with the earnings of an agricultural labourer in the south.

Not expecting to meet Italians on the other side, sections of the Black Flame division immediately surrendered to the Garibaldi fighters. Elsewhere hard fighting took place in a kind of Italian civil war. The Italian Republicans used loudspeakers to broadcast propaganda to the Fascists, 'Brothers, why have you come to a foreign land to murder workers?' Most of the Fascists would not have known how to answer. Republican aircraft dropped leaflets among the blackshirts, promising safe-conducts to all who surrendered, as well as cash incentives to those who brought their guns with them. Blind to all this, Mussolini told the Germans that his troops were mainly encountering Russians.

Roatta tried to combat leftist propaganda by telling his commanders to keep their troops in a state of exaltation. He insisted that the men would fight better if they always had at the front of their minds the idea that the Duce had willed the conflict. In the torrential rain and biting winds it is doubtful if any of the Italians were thinking kindly of the Duce at that moment. With the Italian planes grounded by bad weather that made their makeshift runways unusable, the Fascist troops were kept under constant bombardment by Republican planes operating from permanent air bases. The 3-ton Italian tanks had no answer to the 20-ton Russian tanks used by the Republicans, and their crews abandoned them as useless, even when they could find enough fuel to operate them.

The Italian offensive collapsed in chaos as the Republicans, supported by their Russian advisers, began to encircle Roatta's troops. Before they were completely surrounded the Italians were ordered to retreat. This soon turned into a complete rout. Italian losses were 3,000 killed, 4,000 wounded and 800 taken prisoner. The battle provided a foretaste of deficiencies to be revealed to an even greater extent – and at greater cost – in the next war. The Italians had failed to use their mechanized forces in conjunction with their aircraft and had had no anti-aircaft protection of their own. Mussolini's reaction to the débâcle was to order that no Italian could return alive from Spain unless they had won a victory. The Spanish, both Fascist and Republican, were smiling at the humiliation of the Italians. It was said that after the battle some members of Franco's staff drank a toast to the health of the Republicans, who had shown that Spaniards, even Communist ones, could always get the better of Italians. Conversely, some Italian Fascists were secretly proud that their defeat had been partly at the hands of fellow Italians, even though they were fighting on the wrong side.

CHAPTER 8:
MINISTERIAL IRRESPONSIBILITY

Meddling Ministers

The close relationship between military and political control of the armed forces in the twentieth century is illustrated by a saying usually attributed to Georges Clemenceau: 'War is far too serious a business to be left to the generals.' As grand strategy is concerned with the deployment of a state's military resources to achieve a political aim, political objectives must be related to military capacity. Disaster threatens when the political objective exceeds the military capacity to achieve it, or when military capacity encourages a leader to pursue unrealistic political objectives.

Some have claimed that the British operation to retake the Falkland Islands in 1982 was an example of the first scenario – albeit one which ended in a triumph against the odds. The reoccupation of a group of islands 7,000 miles distant from the UK was, they argue, fraught with risks for a task force depleted by swingeing naval cuts, lacking an aerial early warning system, and vastly outnumbered in the air (see p. 169). A classic example of the second scenario was the Iraqi invasion and annexation of Kuwait in August 1990. Saddam Hussein fatally underestimated the response of the international community to his aggression, and his forces were to suffer ignominious and bloody defeat at the hands of the international coalition ranged against them. Initially pounded by hi-tech aerial power, the Iraqis were then completely outmanoeuvred by more flexible commanders and better motivated troops.

Political interference in military affairs has sometimes led to military disasters on a scale even more vast than this. World War II contained a number of examples of political interference, not only by Adolf Hitler and Benito Mussolini, but by Josef Stalin and Winston Churchill among others. While both Hitler and Churchill were gifted amateur strategists and Stalin a brutal pragmatist, it was Mussolini's strategic dilettantism which inflicted the greatest humiliations on his country, first in North Africa and then in Greece.

The backseat driver

Mussolini's attempted invasion of Egypt in September 1940 was one of his most unfortunate military decisions. Underestimating completely the capacity of the British to respond, and taking the boasts of his generals at face value, he embarked on a fool's errand, the repercussions of which were to shake him from power three years later. In numerical terms Italian forces in Africa seemed impressive: 300,000 in Libya and a further 200,000 in Ethiopia massively outnumbering the tiny British presence in Egypt. But wars are not won on paper; the Italian military machine was inefficient in every way.

In June 1940 the Italian commander-in-chief in Libya, Marshal Balbo, was killed when his plane was shot down by his own anti-aircraft guns. His successor, Marshal

Italo Balbo, a brilliant flier, suffered the ignominious fate of being shot down by his own side while flying to take up command of Italian forces in Libya in 1940.

Graziani, at least realized that the army was inefficient, short of every kind of modern weapon and poorly supplied. But he also knew that a dictator like Mussolini could not swallow a pill as bitter as this and that he would have to try to postpone military operations while attempts were made to remedy the deficiencies. Graziani's first ploy was to exaggerate the strength of his opponents. According to the Italian commander, French and British strength in the region amounted to a formidable 650,000 men against no more than 140,000 Italians. The truth was slightly different. The only troops who could be deployed against the Italians were Wavell's 35,000 British, of whom nearly half were administrative and non-combatant. But if Graziani thought Il Duce would be impressed with the argument that the Italians were vastly outnumbered he was wrong.

Mussolini was determined to win a victory in North Africa for political rather than military reasons. His jealousy of German successes in Norway and France, and the contemptuous way in which Hitler only informed his ally of what he was going to do after he had done it, made Mussolini determined to achieve an impressive victory over a stricken enemy. He ignored Graziani's claims about the size of the Allied forces, and listened instead to those generals who belittled British strength and said that the British lacked drive. Convinced that victory could be easily achieved, he pressed on with the campaign, but the British, once roused, inflicted a series of humiliating defeats on the Italians.

If wars were won by boasting alone, Mussolini's generals would be numbered among the great captains of history. Graziani had told Mussolini that 'When the cannon sounds everything will fall into place automatically.' But when the Italian troops were ordered to advance, or as Graziani picturesquely described it, when the 'hounds chafing at the leash' were released, all of them suddenly found more interesting things to do than chase the British hare. The situation, previously so bright, suddenly became fraught with difficulty. The British, who supposedly lacked drive, were suddenly bursting with it. On the Egyptian border two companies of Italian troops were reported to have been overrun by 300 British armoured cars, more than the British had in the whole of Africa. Italian tank drivers, confronted by the armoured cars, abandoned their vehicles and ran away – or, as the official report put it, 'dispersed'.

Graziani's tale of woe grew longer and sadder by the day. Fighting a stronger battle against Mussolini than he managed against the British, he looked for every chance to delay the assault on Egypt that he feared so much. In his words the war had become transformed into the Italian 'flea' fighting the British 'elephant'. Mussolini was quick to support this notion, claiming that Italian 'meat' (strength and courage) would triumph over British 'iron' (superior equipment and technology). Graziani knew that his equipment was quite inadequate for a modern war. His L.3 light tanks had no sand filters and thus were useless in the desert, while the engines of the Italian aircraft were affected by a similar problem.

'For whatever evil may occur, I, before God and my soldiers, am not responsible.'

Marshall Rodolfo Graziani, 7 September 1940

Mussolini pressed the Italian forces in Libya, under Graziani, to invade Egypt and beat the far weaker British forces there. But Graziani found every possible excuse not to do so. Eventually Mussolini told him to attack or resign and on 7 September 1940 he crossed into Egypt, though not before trying to shift the blame for whatever might happen onto everybody else. Graziani's lack of confidence in himself and his men may have been justified, but reflected years of Italian complacency and ineptitude.

By July 1940, Mussolini's patience was running out. He believed that Hitler would follow the defeat of France with an invasion of Britain, and that an attack on Egypt would therefore be an easy option. With Britain fighting for national survival against the Germans, they could hardly worry about their far-flung garrisons in North Africa. He sent Graziani orders that he must attack by July 15, to synchronize Italian actions with German plans to invade Britain. But Graziani still dragged his heels, complaining that the terrain was unsuitable and the temperature too high. Graziani's diaries present a picture of a man whose mind was not focused on war at all. They record time spent at fashionable lunches, at Fascist ceremonies, and at the Benghazi theatre. Graziani could hardly be expected to interrupt his busy social life for a trifling thing like war. Mussolini now became frantic. If Germany succeeded in invading Britain or if Churchill agreed to a negotiated peace, he would have gained nothing at all from the war – not even a little piece of Egypt.

While the Battle of Britain was raging over the south of England, the Italians were fighting a different kind of war – one in which it was possible for Il Duce to demobilize large parts of his army to help with the harvest. The Germans were astonished at the 'Italian way of war'. One observer wrote from Milan:

> Everyone thinks only of eating, enjoying themselves, making money, and relaying witticisms about the great and powerful. Anyone who gets killed is a jerk . . . He who supplies the troops with cardboard shoes is considered . . . a sort of hero.

But on the subject of Egypt Mussolini was in deadly earnest. 'After twelve months of waiting and preparation it is time to attack the forces that defend Egypt . . . The loss of Egypt will be the *coup de grace* for Great Britain.' It was easy for him to say that in Rome. On the other side of the Mediterranean Graziani was having a bad time. His HQ at Tobruk was being repeatedly bombed by the RAF and he had taken to his shelter, accusing all and sundry of treachery. But by now Mussolini had had enough. Either attack, he told Graziani, or be sacked. Graziani replied by accusing Il Duce of 'a crime of historic proportions' in ordering him to invade Egypt.

On 9 September 1940 the Italian advance began. General Maleti, nicknamed 'the old wolf of the desert', got lost inside Italian territory and almost ran out of water. Planes of the Italian air force were sent out to find him and discovered that his men had had to abandon their original plan, having used up all their fuel going round in circles. When the troops crossed into Egypt their formation was described by one British officer as so chaotic that it resembled a 'birthday party'. Some Italians panicked under fire, officers deserted their troops and had to be driven back to the front line by the curiously nick-named General 'Electric Beard' Bergonzoli. These were the soldiers whom Marshal Badoglio had told Mussolini were 'superior to German troops'.

Having made a symbolic gesture by invading British territory, Graziani's men now began to establish fortified positions at Sidi Barrani. The commander relaxed. At least he had managed to quieten Il Duce. One of his generals, Berti, even went on leave to Italy to have his piles looked at and to comfort his sick mother. Clearly the Italians expected the British to be as cautious as they were themselves. But Graziani, having sown the wind – a very tiny wind admittedly – was now about to reap the whirlwind. Before Berti could return from his mother's side his troops had been attacked and over-whelmed. On 7 December, the British under General Richard O'Connor began a reconnaissance 'in force' which turned into one of the most successful campaigns in military history. In less than eight weeks he shattered the whole of the Italian position in North Africa, capturing 150,000 Italians, 400 tanks and 1,200 guns with just 25,000 men and 275 tanks of his own. The British found it particularly odd that so many of the Italian prisoners were taken prisoner carrying neatly packed suitcases. Graziani wrote to Mussolini insisting that nothing of what was happening was anything to do with him. He was guiltless. The fault rested with those in Italy who had advised Il Duce to force him to advance into Egypt. Everything that had happened was predictable. After all, how could a flea fight an elephant?

Sir Anthony Eden merrily observed, 'Never had so much been surrendered by so many to so few.' Hitler, far from being shocked by the Italian disaster in North Africa, was distinctly amused by it: 'Failure has had the healthy effect of once more compress-ing Italian claims to within the natural boundaries of Italian capabilities.'

Not Getting on with your Friends

Wartime creates strange alliances. The democracies of Britain, France and the USA found themselves the allies of Tsarist Russia in World War I and the Communist Soviet Union in World War Two. On both occasions there were many people – politicians and soldiers – who felt that they were fighting the wrong enemy. And yet the need to defeat Germany in both wars made it possible for democracies and autocracies to overlook their differences in a common struggle for survival. Self-interest is a ruling motive in the lives of states as well as individuals. But there have been times when politicians have, for whatever reason, felt unable to compromise on differences with a natural ally and have accordingly suffered military disaster at the hands of a common enemy.

A Byzantine blunder

The emergence of Islam in the seventh century offered a supreme challenge to the Christians of the Byzantine Empire. The Emperor Heraclius, fresh from his triumphs over the Persians, faced the new threat from the desert with a heavy heart. Years of campaigning had weakened him both physically and mentally, and it is an irony of history that the Arabs should begin their expansion at exactly the point in time when their most formidable opponents, Byzantine and Sassanid, were as exhausted as prize fighters in the last round of their title bout. Triumphant though Heraclius was, he had to face internal doctrinal disputes and this sapped his energy. It may also have affected his judgement, for his conduct of the Byzantine campaign in Syria showed none of the certainty which had been the hallmark of his campaigns from 622 to 629.

During the late sixth and early seventh centuries Christian, Jewish and Zoroastrian missionaries had been at work among the desert tribes of Arabia. Many of these erstwhile idol-worshippers accepted Christianity in its Monophysite form, particularly tribes like the Banu Ghassan, who lived on the Arabian border of the Byzantine Empire. These tribesmen had been used as mercenaries by previous emperors and had formed a useful bulwark against the Persians during the recent Byzantine–Sassanid wars. It would certainly have been in the interests of the military commanders to make sure that such useful allies were well treated. But the Byzantine emperors were not just military leaders. Heraclius had a responsibility as the physical arm of the Orthodox Christian Church and could not easily allow political or military factors to sway his judgements on religious affairs. It happened that just when he needed the military assistance of the Monophysite Arab tribes, they were undergoing a period of persecution by the Orthodox Church. Heraclius himself was deeply involved in this process and was trying to impose on the Banu Ghassan a new doctrine called Monoenergism as a compromise between their Monophysite beliefs and the Orthodox Church. This turned out to be very unpopular and Heraclius's intervention was deeply resented.

All of this coincided with the expansion of the Muslim Arabs, who moved into Palestine threatening Jerusalem. When they captured Gaza in Egypt in 634, Heraclius saw that they represented a very real threat. He responded by sending his brother to Syria with a large army, but he was defeated. A second army under Theodore Trithyrius also suffered defeat. Finally a strong force under an Armenian general, Vahan, part of which consisted of 12,000 Christian Arabs, was despatched against the Muslims. On 20 August

636 the decisive battle of Yarmuk was fought in a blinding sandstorm. During the fighting the Banu Ghassan and other Christian Arabs went over to the enemy, thereby ensuring a Muslim victory. It was one of the most decisive acts of treachery in history. Hopelessly outnumbered after the defection, the Byzantines were overwhelmed and with their defeat the whole of Syria and Palestine – including Jerusalem – fell into the hands of the Muslims. Everywhere the Monophysite Christians welcomed the followers of Muhammad for saving them from the grim orthodoxy of Byzantium. By losing the support of these border tribes Heraclius had lost not just a battle or even a war but the birthplace and home of his religion.

The doctrinal disputes of the seventh century may seem a far cry from the political struggles of the twentieth century, between Fascism and Communism, yet Heraclius's error was repeated by Hitler in 1941 with disastrous consequences for Germany.

Flowers in the barrel of a gun

By 1941 Stalin's regime in the Soviet Union was intensely unpopular and nowhere was this more apparent than among the subject nationalities, most of whom still dreamed of the self-determination which had been granted to many European peoples by the Peace of Paris in 1919. Given the right encouragement and, most of all, protection from Stalin's savage secret police, minorities like the Ukrainians, Latvians, Lithuanians, Armenians and Georgians would probably have taken up arms against the Soviet forces. Hitler had been told by his minister in Moscow, Count Schulenberg, that if he treated the subject peoples in the right way they could become frontline troops in his attempt to defeat the Communists. But Hitler was not prepared to take advice from Schulenberg. For him the war was a crusade and the purity of the crusaders was as important as ultimate victory. To ally with *Untermenschen* or sub-humans, as he regarded all non-Aryans, was anathema.

Yet, as the early days of the German invasion were to show, this was a strategic error. As the Red Army crumbled and the Germans pushed further into European Russia they found that they were welcomed as liberators. Not just peasants but Red Army officers with whole squads of men surrendered willingly, asking only the chance to fight against Stalin. Field Marshal Leeb's Northern Army Group was overwhelmed by the reaction of the Baltic peoples, while the story was repeated in the Ukraine as Cossacks flooded to join the German invaders. Hitler was presented with the opportunity of organizing an anti-Bolshevik crusade. But he missed his chance because he could not bring himself to compromise his racial theory. His immediate plans were to occupy Russian territory and depopulate it, sending its inhabitants back to Germany to work as slave labour. In the long term the new territory was to be opened up to German settlers. Thus was an unparalleled military opportunity wasted because Hitler lacked the flexibility and the foresight to see where Germany's best interests lay.

In the face of Hitler's clumsiness Stalin showed an unusually deft touch, winning back so effectively those who had welcomed the Germans with flowers that they quickly exchanged their bouquets for weapons. Hitler's extermination squads, who followed the front-line troops into Russian territory, played Stalin's game for him. Against an aggressor intent on killing or enslaving whole peoples there was no answer but war to the death. Instead of an anti-Communist crusade organized by Nazis, the eastern front witnessed an anti-Fascist crusade in which many became partisan fighters, determined to kill a German before they were themselves killed. Stalin built the struggle into a national war, scoring a propaganda victory with stories of previous national victories over invaders. Eisenstein's great film of *Alexander Nevsky* showed the Prince of Novgorod triumphing over the Teutonic Knights, who, in spite of their shields and armour, were clearly identifiable with Hitler's twentieth-century Nazis. In concentration camps inside Germany Hitler held five million Soviet prisoners whom he was in the process of working to death. It is estimated that 2–3 million of these would have fought against Stalin in 1941. It was a chance missed. Yet it is doubtful if Hitler would ever have regarded it as a real option, even if his generals had dared to recommend it.

'The duty of the men at Stalingrad is to be dead.'

Adolf Hitler, January 1943

In 1943 General von Paulus's Sixth Army was surrounded at Stalingrad. Rather than allow them to attempt a breakout earlier – or, failing that, surrender – Hitler ordered them to die at their posts. Wars conducted as trials of will between dictators result in a huge wastage of manpower. The Sixth Army could have retreated but to have done so would have been an admission of weakness. Hitler even promoted von Paulus Field Marshal on the grounds that no German of that rank had ever surrendered. But Hitler was flying in the face of reality. The Germans surrendered.

Media Muddles

The role of the media in modern warfare is a sensitive one. In an open society the press must be open to a degree of self-censorship, otherwise the very freedom it enjoys may damage the national interest. Impartiality and a desire to present the truth to readers thousands of miles away from events must be tempered with the knowledge that vital military information can be given to the enemy in apparently harmless press releases. Military authorities have to decide how much press representation is of benefit to national morale. In the area of total war the media becomes a weapon of the state, yet in most British campaigns since 1945 the press, radio and television have been allowed considerable freedom. This has created its own problems. All media forms have developed their own methods of military speculation, frequently involving panels of retired admirals and generals who are called on to give expert opinions. The aim – a laudable one – is to keep the public informed. It is not designed to help the enemy. Yet there have been occasions when this has happened. In the Falklands War of 1982, information in the British press alluding to the failure of Argentinian bombs to explode was very helpful to the enemy (see below), as were reports that SAS and SBS patrols had landed on the Falklands prior to the landings at San Carlos.

In these cases the newspapers, though following their normal policy of keeping the public informed and competing for readership, were disregarding wartime conditions. Some areas of activity are so secret or so vital that they need to be protected from public knowledge. In 1944, had the exact date and location of the allied landings in France been available in the Allied press the consequences would have been catastrophic for the Allied cause. It would have been absurd to risk the lives of servicemen and the interests of the nation in pursuit of a 'scoop'.

An explosive story

In the Falklands War in 1982 one particular revelation by the BBC (from an MOD press release) could possibly have affected the outcome of the war as a whole. Up to 23 May nearly fifty per cent of Argentinian bombs which struck British warships failed to detonate, probably because the low-level attacks by the Argentinian pilots gave the bomb fuses insufficient time to arm themselves. It was important to keep this information from the Argentinians; once they became aware of the problem they would look for a way of overcoming it, either by improving fuses or altering bombing techniques. But on 23 May 1982, a BBC news bulletin carried the following information:

> Following the Argentine air attacks on 21 May two unexploded bombs on one warship have been successfully defused and a further one dealt with on another warship. Repairs are being carried out on the other warships, which sustained minor damage in the raid.

This press release, as used by the BBC, originated from the Ministry of Defence in London. Some servicemen in the Falklands regarded the information as so useful to the enemy that they considered its release as an act of treason. Nor should such a view be rejected lightly. Had the Argentinians managed to overcome their bombing deficiencies many more ships and hundreds, possibly thousands of lives might have been lost. Luckily for the British task force, the Argentinians made no use of the information. When HMS *Plymouth* was hit by four bombs a few days later, any one of which could have sunk her, none exploded. During the entire war as many as 75 per cent of Argentinian bombs failed to explode. Had this figure been reduced to even 50 per cent, Britain's task force might have been irredeemably damaged and the war lost.

A near miss on a British warship in San Carlos Water during the Falklands War of 1982. The British Task Force was spared even greater losses by the failure of a high percentage of Argentinian bombs to explode on hitting their targets.

Cross words for *The Times*

The capacity of the press to exercise a direct influence on military events is not purely a twentieth-century phenomenon. During the nineteenth century *The Times* newspaper had an unparalleled reputation for truthfulness throughout the world. With the British expedition to the Crimea in 1854 went its correspondent W. H. Russell, who sent back to London some of the greatest – and some of the most mischievous – despatches ever sent by a war correspondent. In pursuing the truth Russell frequently felt the need to comment forcefully on the incompetence of the support services as well as the military leadership. Indeed it might not be claiming too much to say that Russell broke the nerve of the British commander, Lord Raglan, and hastened his death.

The problem was that Russell, by making public so much military information, might have been unintentionally helping the Russians. In fact, the evidence suggested that *The Times* actually reached the Russians besieged in Sebastopol before it got to the British camp at Balaclava. On 23 October *The Times* carried an article which gave details of the exact location of the British regiments and of a powder mill. Almost immediately the mill came under Russian bombardment and was blown up. Lord Raglan wanted to know how the Russians could have located the mill without the information in *The Times*. He was furious and wrote to the Secretary of State for War in London, the Duke of Newcastle, 'I am quite satisfied that the object of the writer is simply to satisfy the anxiety and, I may say, the curiosity of the public, and to do what he considers his duty by his employers, and that it has never occurred to him that he is serving more essentially the cause of the Russians.'

Raglan felt that he was not getting enough support from London. He thought that the government should intervene to control the press. Newcastle eventually asked for and got a promise from John Thadeus Delane, editor of *The Times*, that he would be more careful about what he published in future. But what Lord Raglan did not realize was that Newcastle was quite content for the military commanders to take the blame for the mismanagement of the war. Raglan wrote to Newcastle in desperation:

> The paid agent of the Emperor of Russia could not serve his master better than does the correspondent of the paper that has the largest circulation in Europe . . . I am very doubtful whether a British army can long be maintained in presence of a powerful enemy, that enemy having at his command through the English press and from London to his Headquarters by telegraph, every detail that can be required of the numbers, condition and equipment of his opponent's force.

In spite of Raglan's protests nothing was done to censor Russell's despatches and the Russians undoubtedly did benefit from this leakage of information. The Tsar boasted on one occasion, 'We have no need of spies, we have *The Times*.' The ordinary soldiers certainly felt that *The Times* wanted to see them beaten. On one occasion it even published details of potential British weaknesses, notably the thinly defended eastern flank of the army – an invitation for the Russians to attack. In a more cynical age the Russians might have suspected a trap. But in the nineteenth century *The Times* was a guarantee that what you read was what you would find.

THE BATTLE OF ADOWA (1896)

Politicians fight battles from offices and all they spill is ink. The generals they command spill the blood and take the blame. The battle of Adowa was fought for no appreciable strategic reason; it was forced on the Italian general, Baratieri, because Italian Prime Minister Crispi needed a military victory to justify the Italian presence in Abyssinia and to boost his political standing in the country. The fact that Baratieri was an incompetent commander is less important than the fact that he was forced to fight a battle he had no chance of winning.

Italy's rôle in the 'Scramble for Africa' in the late nineteenth century was inspired not by economic or strategic reasons but by the need to match the achievements of 'great powers' like Britain, France and Germany. Italian colonial policies did not evolve from the internal needs of the mother country but were artificially imposed on a state that had neither the money nor the military skill to maintain them. In a sense the disaster at Adowa was inevitable, given Italian attitudes and abilities.

If Italy was the weakest of the European colonial powers, fate played an unkind trick by matching her against the most formidable native military power in Africa. The Ethiopians were able to raise armies of over 100,000 men and from the 1880s onwards they were increasingly armed with modern rifles and even artillery, bought from France and Russia among others. With their knowledge of the terrain and the hard military experiences of war against the Sudanese, the troops of Ras Menelik of Shoa were more than a match for the conscript troops Italy sent to Africa. The Italians also committed the cardinal sin of underestimating their opponents. Perhaps they should have taken a lesson from the British, who had learned to their cost over the centuries not to underestimate their colonial foes.

Fifteen years of occupation had brought Italy high costs in expenditure and military disaster. Crispi told Baratieri that he held in his hands the honour and dignity of the country. Crispi had, in fact, become dissatisfied with Baratieri's lack of action and had ordered his replacement, General

Baldissera, to sail for Eritrea even as Baratieri took the fateful decision to advance on Adowa. One has to sympathize with Crispi's exasperation – he had sent Baratieri as many troops as he had requested, plus more generals than he could use. Whether Baratieri knew he had been sacked and sought battle to try to regain favour will never be known. All that can be said with conviction is that no general was ever put under greater political pressure. Just before the battle, Crispi wrote to him:

> This is a military consumption, not a war: little skirmishes in which we are outnumbered by the enemy; gallantry wasted without results. I have no advice to give you because I am not the man on the spot, but it appears to me that there is no fundamental conception underlying the campaign and I want that remedied. We are ready for any sacrifice to save the honour of the army and the prestige of the crown.

Baratieri's 20,000 troops were dependent for supplies on a 100-mile line to Massawa on the coast and by the third week in February 1896 they were close to running out of food. To bring more from the coast for such an army required 9,000 camels; Baratieri had just 1,700. Resupply was therefore impossible. If a battle was going to be fought it would have to be soon, otherwise retreat – and national disgrace – would be inevitable. Baratieri was under such pressure that his health began to give way and his mental grasp of the campaign was lost. Anxiety had reduced him to a kind of fatalism: he feared defeat but was somehow drawn towards it like a moth to a flame. On 28 February, he assembled his senior commanders and gave them the choice – fight or retreat. Emotions ran high and retreat was equated with national humiliation. The three generals fresh from Italy were all in favour of fighting, though none had experience of campaigning in Africa and all were overconfident. Finally the decision was taken to advance towards Menelik's camp in the hope of drawing the Ethiopians into a general battle.

As in so many colonial disasters, faulty maps played a significant part in the battle of Adowa, as did the rashness and stupidity of European commanders fighting foes they mistakenly believed to be inferior. Baratieri divided his army into four brigades and decided to advance on Menelik's camp, having been misled by the reports of spies that half of the Ethiopian army was off scouting for food. The crucial blunder was made in Baratieri's instructions to Albertone, the most aggressive of the Italian generals and the commander of the left flank of the Italian army. Albertone was ordered to advance to a hill marked Chidane Meret on the commander's sketch map. The map was in error, the hill he had marked was in fact five miles beyond the real Chidane Meret (the point where Baratieri intended the rendezvous to take place). On the right, Dabormida was to advance towards Rebbi Arieni, while in the centre Arimondi led a forward brigade and Ellena a reserve brigade towards Rebbi Arieni.

The night march proposed by Baratieri degenerated into chaos. Though they had the advantage of bright moonlight to march by the Italians behaved as if they were playing blind man's buff. Albertone marched westwards instead of north and collided with Arimondi's brigade. Arimondi had to call a halt to let the left flank brigade pass through and by the time he restarted he was so far behind that Albertone was cut off from the rest of the army. At 5.30 am he arrived at a place he assumed was Chidane Meret, but was of course five miles further on from where Baratieri expected him to be. Albertone immediately attacked some of Menelik's skirmishers, thus bringing on an action while in the wrong position. Hopelessly exposed, with both flanks open, Albertone sent messages back to Baratieri asking for help, but none got through. Soon 30,000 Ethiopians had ringed him round and had occupied the high ground overlooking his doomed command. Outnumbered seven to one it was only a matter of time before he was overwhelmed. The Ethiopian artillery, reputedly controlled by the Empress Taitu herself, devastated his native troops. Fresh Ethiopians poured into the attack and when ammunition ran short Albertone's men were overrun and massacred, although the general himself was taken alive.

The battle took the form of three isolated encounters. While Albertone's brigade was being wiped out on the left, Arimondi's central column was hit by a deluge of Ethiopian horsemen. Its broken remnants mingled with the reserve brigade of General Ellena. Soon, like Albertone, Arimondi's men were forced into a defensive circle and surrounded by the vastly more numerous Ethiopians. The folly of the Italians in not dressing their officers in the same uniform as their men was now revealed, as the white-uniformed officers with

sashes and red and black insignia were the first targets for the enemy sharpshooters. Baratieri gingerly ordered a tactical withdrawal, but this soon turned into a full retreat, with Arimondi left dead on the field.

In splendid isolation Dabormida's brigade, who could hear fighting in the distance, but were incapable of offering assistance to their colleagues, blundered about in the hills on the right of Arimondi. In the morning Dabormida drove off an attack by Shoan forces and, unaware of the disasters which had struck the other Italian brigades, was feeling quite satisfied with himself. However, his turn was coming. During the afternoon some 50,000 Ethiopians surrounded the 4,000 Italian troops and after a brief struggle overwhelmed them, Dabormida being killed in the rout.

Italian losses were enormous: 6,000 men died, 3,000 were taken prisoner and only 1,500 escaped – in a wounded condition. The reason for the high proportion of deaths was that the Ethiopians set fire to the grass on the hills, flushing out those hiding and burning others to death. Menelik's losses were even higher than the Italians but formed a smaller percentage of his total army.

OPERATION BARBAROSSA (1941)

Hitler's decision to invade the Soviet Union in June 1941 was based on his belief that the Soviet system was ripe for demolition. 'We have only to kick in the door and the whole rotten structure will come crashing down,' was how he described the thinking behind Operation Barbarossa. There was much evidence to support his view, notably the purge of the Red Army by Stalin in 1938–9 and the poor performance of the Red Army in the 'Winter War' against Finland in 1940 (see p. 94). But Hitler was fatally underestimating the military potential of the Soviet Union and the political strength of Stalin's regime. He had used the Nazi-Soviet Pact to neutralize the Soviet Union in 1939 so that he could concentrate on war against the Western democracies. But in 1941 he had not achieved his aim of defeating Britain. To plan an escalation of the war even for reasons of grand strategy – like the search for *lebensraum* in eastern Europe and the

German armoured forces enter a devastated suburb of Stalingrad during 'Operation Barbarossa'.
Despite crushing early successes the overstretched Germans were unable to maintain their forces on a
vast eastern front stretching from the Baltic to the Black Sea.

aim of securing the oilfields of the Caucasus – while he still faced an intransigent enemy in the West and in North Africa was the height of folly.

Even before he had lost the air war over Britain in 1940, Hitler had decided that Russia was to be the next target. The planning of Operation Barbarossa was initially carried out by General Erich Marcks, who produced a blueprint in August 1940. Marcks laid great emphasis on defining objectives. For him the principal objective was the destruction of the centres of Russia's war economy around Leningrad in the north, around Moscow in the centre, and the Ukraine in the south. Of these three, Moscow – the economic, political and spiritual centre of the USSR – was the prime target. It was hoped that its capture would fracture the co-ordination of the Soviet war effort. Even though this strategy had not worked for Napoleon in 1812, Marcks was convinced it would work for Germany in 1941. This plan was accepted by the German General Staff and tried out in a series of war games. Whatever modifications German planners called for, no one questioned Marcks's central thesis that the offensive against Moscow was crucial to the success of the operation – and the war.

Hitler, however, did not agree with the General Staff. He had already asked General Jodl at Army Central Command (OKW) to arrange a separate study by General Baron von Lossberg. It was Lossberg's plan which was to form the basis for 'Operation Barbarossa'. Lossberg rejected the notion of a strike against Moscow, and concentrated instead on the Smolensk region to the west of the capital. Inroads were to be made by flanking army groups both in the north and the south while in the centre the Red Army would be 'rounded up' near Smolensk. Believing that 'Moscow is of no general importance', Hitler rejected Marcks's plan in favour of Lossberg's. In spite of months of careful planning by the best brains in the German army, he simply swept aside their findings. Hitler now began to play an active part in planning the operation. He involved himself in specialist areas like logistics, divisional command and bombing tactics, selecting targets in a disjointed fashion. As one historian has written, he was about to send the German army into the Soviet Union 'on a four-year will-o'-the-wisp chase after seaports, cities, oil, corn, nickel, manganese and iron ore'. He even rejected the intelligence report of his Moscow ambassador, Count Schulenberg, filing it in a draw without looking at it and telling the count, 'Thank you, this was extremely interesting.' Hitler generally suppressed intelligence reports which clashed with his own preconceptions. As a result, the German forces remained ignorant of the quality of the Soviet T-34 tank – a weapon which was to have an unexpectedly large impact on the campaign.

British military experts believed the Germans would defeat the Red Army easily. But there was deep uncertainty about what would happen next. The war was expected to enter a new political phase – or in Nazi terms a politico-racial phase – in which vast areas of European Russia would be occupied and colonized. But if Russia refused to admit defeat after her initial disaster and Stalin continued fighting from beyond the Urals, how could the Germans possibly conduct a campaign that pushed ever further into Asiatic Russia? So many troops would be committed to maintaining the ever lengthening supply lines that he would not have enough men left to maintain fighting forces on other fronts like North Africa or potential fronts like northwest Europe. And if the Americans entered the war how could Germany stop them building up their strength in Britain before opening a second front in France? To questions like this Hitler had no answer. His belief in *blitzkrieg* failed to take account of the sort of elemental struggle that geography imposed on any invader of Russia, in which terrain and weather were added to the human enemies.

Hitler's armed forces presented an impressive picture in 1941. They were about to undertake the greatest campaign in human history; even the name 'Barbarossa' had a Wagnerian ring to it. However, Hitler may have been blinded by previous successes in Poland and France. The Panzer divisions of 1940 had 300 tanks each, those of 1941 had less than 200. Of the 4,198 German tanks available 1,156 were still the largely obsolete Mk Is and IIs, while 772 were of Czech Skoda make. Only 1,400 of the modern Mk IIIs and IVs were available, reflecting a poor level of output by German industry. The Russians, on the other hand, fielded 22,000 tanks – many antiquated – and lost 17,000 in the first five months of fighting. But as the war dragged on, newly built Russian tanks – T-34s – proved superior to the latest German ones. Nor was the great campaign as mechanized as Hitler had supposed. Alongside the 600,000 lorries and other motor vehicles, there would be 625,000 horses used for transport, many pulling simple country carts.

The great imponderable in June 1941 was why Hitler decided not to move on Moscow with all speed before the weather made transportation so difficult. Had the capital fallen, Stalin might possibly have lost his nerve and sued for peace. Hitler was a student of history but no historian. His methods were intuitive rather than empirical and he took from the past only the lessons that suited him. He studied the campaigns of Napoleon and Charles XII of Sweden and noted the reasons for their failures. Yet for a comparative model he really needed to look no further than 1914–17. In World War I, in spite of suffering unprecedented losses in the titanic defeats in East Prussia and Galicia in 1914–15, Russia had shown great resilience by fighting back during Brusilov's offensive of 1916. It was a grave mistake on Hitler's part to underestimate the recuperative powers of the Russian people. The Communists were able to turn the fight into a national crusade on a scale similar to that of 1812. Admittedly this had not been possible in 1917, but the collapse suffered by Russia on that occasion was of a political rather than a military nature. Hitler should perhaps have asked himself whether a stronger political system than the tottering regime of Nicholas II might be able to mobilize the full potential of the Soviet Union. But Hitler remained unconvinced, declaring on 4 July 1941 that the war was over and the Russian army and air force smashed. Admittedly, the Soviet Union had suffered losses on a scale never seen before in warfare, yet the thousands of tanks and planes lost were antiquated and could soon be replaced by more and better machines, while the loss of soldiers and inefficient generals cleared the way for a new Red Army under men like Marshal Georgi Zhukov. In addition, Hitler failed to realize that the 15 per cent of troops and 25 per cent of tanks that Germany had lost could not be so easily replaced. In time, the Soviet Union could only get stronger, Germany weaker.

And the key question remained unanswered. What was Hitler's absolute and fixed purpose to which everything else was subordinate? Most of his independent-minded generals, men like Brauschitsch, Halder, Bock and Guderian, favoured a strike at Moscow. Yet Hitler upbraided them, 'Only completely ossified brains, absorbed in the ideas of past centuries, could see any worthwhile objective in taking the capital.' But they could not pin him down to a firm alternative. His butterfly mind flitted from one idea to another. Sometimes he favoured pressure in the north on Leningrad, sometimes in the south against the Donets Basin, the Crimea and the Caucasus. While the flanks were ordered to drive forward, the central group was told to cut off the capital but not to enter Moscow. Apparently the object of this was to deprive the Russians of 'the government, armaments and traffic centres around Moscow', and so prevent a rebuilding of the shattered Soviet armies. But this was a serious mistake; Stalin simply pulled back behind the Urals and rebuilt there, free of German interference. The fighting around Moscow served the same principle as Kutuzov's scorched earth policy in 1812: it gave the Soviet Union time to regroup.

One of the guiding principles in war is singleness of purpose and concentration of available forces to achieve it. Generals know this, but meddling, if gifted, amateurs like Hitler do not. Had he allowed his generals to take Moscow before the weather intervened who knows what consequences might have followed? Instead, he frittered away his resources, widening the battle front against a numerically superior enemy. He also failed to come up with a contingency plan when the predicted collapse of the 'rotten structure' of the Soviet Union did not materialize.

In his contempt for the Russian army – men, officers, equipment and strategists – Hitler believed he could win a *blitzkrieg* without having to consider the problems of fighting a campaign in a Russian winter. Recent history provided the evidence of what these problems were. Many of his senior commanders had served on the Eastern Front in World War I and knew how an army could disintegrate in the harsh conditions. But Hitler rejected logic, even common sense, insisting instead that willpower could overcome any hardships. This was simply metaphysical nonsense to the troops who were freezing in icy wastes because they had not been equipped for winter campaigning. By November and December 1941 millions of German troops were suffering because inadequate forethought had gone into planning the operation. Guderian tried to spell out to Hitler the stark differences between the thinly clad German soldiers and the well-fed, warmly clothed Siberian troops, who had begun to arrive at the front fully equipped for winter campaigning.

The offensive had to be halted and on 8 December even Hitler acknowledged that the Germans would have to stand on the defensive. When his generals recommended a strategic withdrawal to more defensible positions and more comfortable billets for the men, the Führer was adamant: not a metre of ground was to be conceded. Willpower was everything: to retreat one step would indicate a weakening will.

It is doubtful if a politician has ever interfered so decisively in any military campaign as Hitler did in the invasion of Russia. The World War I corporal did not really respect his generals. He believed that his (shallow) reading of military history and his occasional inspired hunches raised him to the level of a great commander, far above the journeyman efforts of Manstein, Rundstedt and Guderian. He confided to Halder, 'Anyone can do the little job of directing the operations in war.' During the winter of 1941–2 he purged the army of commanders whom he claimed had lost faith. The list of victims was impressive, including Rundstedt, Bock, Leeb, Hoepner and Guderian. Von Reichnau, a physical fitness fanatic, was lost as well when he dropped dead from a heart attack after a cross-country run in sub-zero temperatures.

The removal of so many able generals seemed to act as a stimulus to Hitler's own confidence in his mastery of the art and science of war. The famous Prussian military machine, created in the seventeenth century, in which meticulous planning had gone hand-in-hand with the professional judgement of highly trained staff officers, was replaced by 'willpower and ideological fervour'. Knowledge and reason counted for nothing in Hitler's eyes, only the unbending will to win. Generals were no longer free to act on their own initiative; all critical decisions were made by Hitler as supreme warlord. Statistics were challenged or rejected if they contradicted his views. The strains of wartime led to bouts of savage fury. When presented with irrefutable evidence that Russian tank output would reach 1,200 per month, Hitler foamed at the mouth and struck the officer reading the report.

The failure of willpower was most starkly seen during the terrible battle of Stalingrad, which marked the turning of the tide in the war in the East. Hitler forbade surrender and promoted Sixth Army commander Friedrich von Paulus to field marshal on the grounds that no German of that rank had ever surrendered. Promises that the Luftwaffe would keep his men supplied were of no use to the besieged commander. In the end he surrendered because there was no alternative. Willpower could not triumph over the military logic of inferior manpower, equipment and supplies. Unlike any previous military commander, Hitler was contemptuous of his own generals. This reflected his own personality defects as well as the jealousy of the amateur strategist for the professional soldier. The outcome was that few, other than sycophants, kept his favour for long. In May 1943 Goebbels admitted that Hitler was sick of his generals, though one imagines no sicker than they were of him. He had taken them into a war that they could not win and had thwarted their efforts to fight a realistic campaign in the Soviet Union. Germany's defeat on the Eastern Front was less a defeat for the Wehrmacht and its commanders than for Hitler himself. Rarely has any military campaign and its outcome been so much the product of one man.

THE BAY OF PIGS (1961)

After the takeover of Cuba by Fidel Castro's revolutionaries many Cubans found refuge in the United States. By the end of 1959 Miami was a centre of anti-Castro political activity and it was here that the exiles planned a return to their homeland to overthrow the Cuban leader. They were encouraged by officials within the Eisenhower administration, who feared Cuba's Communist leanings – especially in an island so near to the American mainland. From the interaction of these two interests the military fiasco at the Bay of Pigs was born.

Early in 1960 President Eisenhower authorized the arming and training of a Cuban exile 'army of liberation' under the direction of the CIA at secret bases in Guatemala. The aim was to prepare an invasion of Cuba by an armed force of Cuban refugees. Their landing would be assisted by sixteen old B-26 bombers supplied by the United States but flown by Cuban exiles. It was not officially intended that the Americans should intervene directly, but CIA advisers apparently led the Cubans to believe that their landing was just a prelude to further American involvement.

When the presidential election of 1960 brought John F. Kennedy to the White House he inherited a plan that was already some way towards fruition and which he would find difficult to stop. He was clearly unhappy at inheriting from Eisenhower the advisers who had planned the strategy, but, inexperienced as he was, he let them have their way.

The CIA plan had developed from an original idea of guerrilla infiltration into a proposal for a full-scale amphibious operation, designed to land a substantial force of exiles. Air strikes would be carried out by the B-26s from landing fields in Nicaragua to knock out Castro's air force, and would continue to support the actual landings. The troops would be supplied with artillery and would expect to take control of a large area on the coast to enable anti-Castro elements in the country to rally to them. At this point a provisional government would be established, recognized by the USA and then openly supplied with logistic if not military aid.

As a military plan this had obvious flaws. An evaluation by the Joint Chiefs of Staff stated that the success of the operation depended on either a sizeable uprising inside Cuba or else on substantial support from outside. Yet this part of the report was immediately contradicted by the extraordinary assertion that the plan stood a 'fair' chance of success anyway and would still contribute to the overthrow of Castro's regime. Clearly someone was guilty of wishful thinking, always a dangerous basis for military planning.

The political implications of the report were very important. In the first place, Kennedy made it clear that there must be no direct American military intervention, so the possibility of outside support had to be disregarded from the outset. However, the CIA and, through them, the Cuban exiles had convinced themselves that the Americans would not allow the invasion to fail and thereby risk national humiliation. They therefore interpreted the President's statements rather loosely.

If outside aid was not going to bring about success, how likely was it that there would be a national uprising inside Cuba? The answer was that the Americans could not tell, for the simple reason that they had not tried to find out. They had not contacted any groups inside Cuba and were content to take the claims of the exiles that they would have widespread support at face value. In the event this was a serious miscalculation. In the final analysis the JCS report was indefensible. How could experienced military planners believe that an invasion force of 1,400 Cuban exiles, few of them regular soldiers, had a 'fair chance' of overcoming Castro's army and militia of 200,000?

Members of Castro's local defence force at the Bay of Pigs examine the bloodstained clothing of one of the invaders killed in the fighting. The bungled operation was based on faulty CIA intelligence.

Although cancellation of the entire plan was an option, the new administration was won over by the ardent advocacy of CIA men like Allen Dulles and Richard M. Bissell, Jr, who argued that if the plan was abandoned there would be the problem of what to do with the large force of Cubans presently training in Guatemala. Apparently Kennedy was stuck for an answer. Having been trained to believe that they were going to liberate their country the Cubans might react badly to hearing that the whole plan was cancelled. Once they began to spread the word that the US government had been

planning an expedition against Castro but had lost its nerve, American prestige – and that of the fledgling president, would be seriously impaired.

Kennedy told Dulles and Bissell that he wanted the whole matter kept as low-key as possible. Under no circumstances was there to be any question of American military intervention. But the CIA was not being absolutely straight with the administration. They never really believed that there would be substantial uprisings behind the lines in Cuba, and clearly thought the invasion could only succeed with American military backing. Rather than let the whole thing fail, they believed any American president would have to intervene.

No attempt was made to verify the claims by Dulles and Bissell that there were 2,500 men in resistance groups inside Cuba, that 20,000 more were open sympathizers and that the invasion force would win the support of at least a quarter of Cuba's population. Where did they get these figures? They claimed that 'contacts' in Cuba had informed them that these men were ready to fight when the signal was given. But these claims were inflated 'out of hope or despair' and were the flimsiest possible evidence on which to base a military operation.

Dulles and Bissell had always claimed that if the invasion failed the invaders would have no difficulty in escaping into the mountains to continue a guerrilla war. But this explanation was invalid for two reasons: none of the men had been given any guerrilla training, and the mountains in question – the Escambray range – lay eighty miles from the Bay of Pigs across a difficult terrain of swamps and jungles. In any case, no one actually explained this option to the Cubans themselves for fear of lowering their morale. CIA agents merely told them, 'If you fail, we go in.' Yet Kennedy's decision to let the expedition take place was mainly based on CIA assurances that in the event of failure the Cubans would be able to escape into the hills.

The real problem for Kennedy was that he and his new administration were saddled with someone else's plan – and the planning had now gone far too far to cancel. Eisenhower's administration had created a monster which it could not destroy without immense loss of prestige. Kennedy was innocent of its creation, but as president had the problem of finding a use for it. Dulles's arguments about the difficulty of calling off the operation had convinced Kennedy that the Cuban refugees were less danger to him in Cuba than they were anywhere else, particularly if they were free to give details of their training by American agents. Kennedy felt it was better to 'dump' the exiles in Cuba. Then, if things went wrong, they would not be able to disclose American involvement in the invasion. After all, dead men tell no tales.

D-day was set for 17 April, and the invasion force assembling in Guatemala now numbered 1,400 men. Even at this late stage Kennedy was plagued by doubts. He asked a veteran Marine colonel to inspect the Brigade and report back. The report was immensely reassuring – and totally misleading.

My observations have increased my confidence in the ability of this force to accomplish not only initial combat missions, but also the ultimate objective, the overthrow of Castro. The brigade and battalion commanders now know all details of the plan and are enthusiastic. These officers are young, vigorous, intelligent and motivated by a fanatical urge to begin battle. They say they know their own people and believe that after they have inflicted one serious defeat upon the opposition forces, the latter will melt away from Castro . . .

This incredible document served to misrepresent the true facts as completely as possible. No consideration was given to the strength of Castro's army and air force, which were simply dismissed as insignificant: it was said that the former would be bound to melt away at the first sight of 1,400 'desperadoes' while the latter would be bombed on the ground by obsolescent American planes. The CIA was not being honest either with the President or with the Cuban exiles. Even at the moment of departure, and with Kennedy's assurances of no American military involvement, CIA advisers were telling the Brigade that they would not be the only unit in the landing and that within 72 hours they could count on US military assistance.

If everyone was underestimating Castro's military strength in Cuba they were similarly setting too high a value on the morale of the Cuban Brigade. Brave and determined they might be, but they numbered just 1,400 men, of whom only 135 had experience as soldiers. The rest were students, businessmen, professional people, peasants and fishermen. There were many boys of 16 or under, while one man was over 61. It was not a force to give Castro sleepless nights.

The main question remaining was the part to be played by the B-26s in destroying Castro's air force on the ground. The American State Department felt that any air strikes before the landing had taken place would suggest American collusion. On the other hand, the Pentagon pointed out that without air strikes the landing might not be a success anyway and that Castro's air force needed to be eliminated to protect the disembarkation. A compromise was reached when a bombing strike was organized involving B-26s flown by Cuban pilots pretending to be deserters from the Cuban air force. The CIA estimated that Castro's air strength consisted of fifteen B-26s and ten Sea Furies. Unfortunately, they overlooked four T-33 jet trainers. According to the CIA, Cuba's air force was 'entirely disorganized . . . for the most part obsolete and inoperative'. It was claimed that the raid by the eight US B-26s piloted by Cubans had inflicted total destruction on Castro's air force, but the next day it was clear that just five planes had been hit.

On 14 April the invasion force sailed from Nicaragua in five small freighters. The captains of the invasion ships were uneasy. A CIA adviser gave one a packet of over a hundred nautical charts in a sealed envelope, and told him not to open it until he was under way for Cuba. When he did open it he found two vital charts missing. Aboard another boat, the second officer decided to test the machine gun, which had been badly mounted in Nicaragua. As he opened fire, the barrel sagged and sprayed bullets into the crew, killing one and wounding several others.

The invasion fleet reached its destination on 17 April and prepared to land near the resort town of Playa Giron. The CIA had forgotten to warn them that this beach had dangerous coral just offshore, unlike the sandy beaches where they had practised their landings. The result was that two landing craft had their keels ripped open and their men pitched into the sea. The invaders had been told that Castro's air force was out of action. But as they landed they received a message from Miami warning them that some Cuban aircraft were still intact. In fact, a dawn air strike had been planned for the next morning to destroy any Cuban planes that had survived the earlier strikes, but the President had personally cancelled it. From this moment the operation was doomed.

Once ashore the invaders discovered that CIA planning had let them down even further. They had been told there was no radio transmitter within miles of the beach. In fact, there was one only a few hundred yards away. Warning messages were already reaching Castro as the invaders waded through the surf.

At dawn, Castro's air force struck back with unexpected strength. A Sea Fury sank the ship carrying the reserve ammunition, which, foolishly, had been loaded alongside the communications equipment. A roar, a puff of smoke, and the expedition was reduced to chaos. While the slow-moving B-26s flew defensive missions overhead, they were attacked by the unconsidered T-33 jet trainers, which were armed with machine guns and shot down four of them.

The first air strike had warned Castro of impending trouble and he had responded with great speed, rounding up 200,000 suspected counter-revolutionaries in Havana. At the same time, 20,000 government troops with artillery and tanks closed in on the invaders.

Everything began to go wrong. Castro's military strength had been underrated and there was no mass rising of the population. The invasion was in trouble unless the Americans intervened directly. Bissell pressed Kennedy to allow a concealed American air strike from the aircraft carrier *Essex*, lying off the Cuban coast. The aim was to knock out the jet trainers and allow the rebels' B-26s to attack Castro's tanks. Kennedy compromised by permitting six unmarked jets from *Essex* to fly over the Bay of Pigs and protect the Brigade's B-26s from air attack. However, when the idea was put to the Cuban pilots they said they were too exhausted, and the B-26s were flown instead by American pilots under contract to the CIA. What followed was a blend of tragedy and farce. By an elementary mix-up no allowance had been made for the different time zones between Nicaragua and Cuba, so that the B-26s arrived over the beaches an hour before their American jet fighter cover. The result was that the B-26s came under heavy fire, and four Americans were killed.

Less than sixty-four hours after landing at the Bay of Pigs the remaining invaders were overrun by Cuban troops and taken to Havana as prisoners. Their casualties had been heavy: 114 killed and 1,189 captured. Until the very last moment they had clung to the belief that the Americans would not allow them to be defeated.

In the aftermath of the Bay of Pigs fiasco, Kennedy made every effort to find out how it had happened. How had a responsible government allowed itself to become involved in such a ridiculous and ill-fated adventure?

In the final analysis the operation failed because Kennedy had made the cardinal error of trying to reduce the political risks involved so much that the military risks became overwhelming. The final decision to go ahead was the President's, but his heart was never in it. Later, a shaken Kennedy was to comment, 'All my life I've known better than to depend on the experts. How could I have been so stupid as to let them go ahead?'

The Bay of Pigs fiasco was a grave blow to American prestige and most particularly to that of the new president. It served to strengthen ties between Cuba and the Soviet Union and to persuade the Soviet leader – Nikita Khrushchev – that President Kennedy was a weak incompetent. Only by taking the world to the brink of nuclear war over the Cuban Missile crisis was Kennedy able to convince the Soviets that they were mistaken.

THE BATTLE OF NAMKA CHU
(1962)

In 1962 war came to the disputed territory on India's northeastern frontier with China, high in the Himalayas. Over a number of years India had scattered frontier posts across the disputed territory, but India's leaders were complacent about any possible Chinese threat. Most of the Indian frontier posts were manned by a few poorly armed troops, hardly a match for the well-armed Chinese frontier guards. To the Chinese, India's frontier posts were an invasion of China's territory and were much resented. In September 1962, Chinese troops occupied one of these outposts, using minimal force. They then withdrew, having made the point. They could clear out the Indian invaders any time they liked.

The Indian government did not see it that way. They had a completely false image of Indian military strength in the area and their military advisers were quite unwilling to give an accurate assessment. In fact, the Indian position was hopeless from the start. Whereas the Chinese side of the frontier region had good communications, all-weather roads, and comfortable living conditions for their troops, the Indian side had no modern communications. The Indian troops could only be supplied by air and had to endure appalling conditions of heavy snow and high altitudes. The hardy Chinese troops were chosen from the native population and were well used to the altitude, while the

Indians, often conscripts from the plains, needed weeks of slow acclimatization.

Brigadier J. P. Dalvi, who was sent to command the 7th Infantry Brigade (one of only two brigades in the disputed territory), said that his men had not carried out a single manoeuvre since 1959. Instead of training they spent their time as labourers, building a helipad in the mountains or collecting wood at 14,000 feet to build their own shelters. The best of the non-commissioned and junior officers had been creamed off to help training in other parts of India, and there was a constant shortage of officers and specialists. So severe were conditions where these men were forced to live that the normal complement of a brigade, 3,500, had been reduced to 2,400 because of difficulties in supplying them with food by air. The remaining troops were kept lower down on the plains. In the event of attack these troops would have had to climb the mountains on foot, as wheeled vehicles could not make the ascent. Even after they had arrived, exhausted and quite unfit for combat, they would not have survived without a period of acclimatization – six weeks being the recommended minimum. The outcome was that when the attack came Dalvi's 7th Brigade was at barely sixty per cent strength.

To justify this inadequate situation, Indian military intelligence had used flawed logic to reach a curious conclusion. They reasoned that if they could only manage to maintain a brigade in the harsh conditions, the Chinese could only maintain the same strength themselves. In reality this was quite wrong, yet it was important for the planners in Delhi to believe it, otherwise they could hardly answer the complaints from their front-line commanders. In fact, the Chinese maintained two divisions facing the 7th Brigade.

As the political crisis deepened no attempt was made to advise the Indian government on the weakness of the military position. The army kept quiet while the politicians issued warlike threats. Local commanders in the Northeast Frontier Agency bombarded their headquarters with warnings about Indian weakness and the danger of massive Chinese retaliation. However, their views were ignored and military intelligence blithely advised the government that China would never risk a major war against India. The chief exponent of this theory was the Chief of Staff, General B. M. Kaul. Kaul was actually sent to command the troops in the disputed territory, even though he was a staff officer and had no combat command experience. It seems that Prime Minister Nehru and Kaul were united in believing that China would not dare to take strong action against India.

Meanwhile, the only way to supply Dalvi's advanced forces was by parachuting in food and equipment from the air. This method was highly unreliable, with 75 per cent of loads landing in areas held by the Chinese. Of those that landed in Indian territory, 30–40 per cent were ruined as the parachutes had ripped or failed to open. This failure rate was the result of retrieving and re-using old parachutes, some of which had been lying for months in the open and had rotted. Oil was dropped in 44-gallon drums, so heavy that no one could retrieve them from the heavy snow or haul them up the steep gradients. The boots supplied to Dalvi's forces only came in two sizes – 6 and 12 – and were without studs. Their rubber soles and toe-caps perished in the harsh conditions.

As the Chinese troops seemed unwilling to evacuate their positions, Nehru made the public announcement that they must be 'thrown out', and with these instructions Kaul was sent north. When he arrived at the front, Kaul insisted on immediate action, even though the local commanders tried to convince him that it would be disastrous. Accused of stalling by his political bosses, Kaul had no option but to order an advance on 10 October, the last date permitted in his instructions from Nehru in Delhi. The Indian troops were to be pushed forward, whatever the outcome. In appalling conditions, many Indian soldiers immediately succumbed to pulmonary disorders and frostbite. None had appropriate clothing and all were equipped with hard rations, quite inadequate in the freezing weather. No arrangements existed for evacuating casualties or resupplying the troops with food or weapons. 7th Infantry Brigade was being asked to operate in isolation – in defiance of all known military principles. When Dalvi asked Kaul how his troops could survive in such conditions, Kaul told him that 6,000 sets of winter clothing had been ordered from Canada.

On 10 October an Indian force of 50 men was ordered to move past the Chinese troops and 'sit behind them'. The Chinese, 800 strong, reacted by wiping them out. Kaul was shocked. 'Oh, my God,' he commented. 'You are right, they mean business.' Dalvi pressed Kaul to evacuate the bri-

gade, which was by now facing a full division. Afraid to take such a decision, Kaul returned to Delhi for further advice, completely abdicating his role as a military commander.

At the height of the crisis, with thousands of Chinese troops overrunning his positions, Dalvi received an irate message from the Educational Branch of Divisional HQ, reprimanding him for failing to send a team of athletes to compete in the Sidi Barrani Celebrations. To add insult to injury, he next received a complaint that his troops were not collecting a high enough percentage of air-drops. He was ordered to detail 200 men a day to collect the loads.

In spite of the seriousness of the crisis, in Delhi all was calm. The next day the Prime Minister left for an official visit to Ceylon (now Sri Lanka), telling the press that the army had clear instructions to drive the Chinese from Indian soil.

Reserve troops, arriving on the Thag-la Ridge, were equipped with just pouch ammunition and rifles; there were no automatic weapons. Exhausted from their climb, these men soon succumbed to altitude sickness. Without machetes, axes or other tools they tried to cut trees for shelters with shovels, much to the amusement of the watching and well-supplied Chinese. Without any food or winter clothing, these support troops merely served to weaken Dalvi's position by diminishing his limited stocks of food and clothing. In the absence of an adequate medical station, casualties had to be evacuated by air; but the only available helicopter was a two-seater.

On 18 October General Kaul's health collapsed and he was taken back to Delhi with pulmonary troubles, no doubt brought on by the high altitude. He left behind no deputy and tried to conduct the campaign, as it developed, by telephone from his hospital bed – a thousand miles away from the front.

By this time, the Chinese had decided that India needed some more tangible demonstration of the hopelessness of their position. Nehru's talk of 'throwing them out' clearly rankled. As one Chinese general said, 'The Americans cannot throw us out, what can you miserable Indians do to us?' On 20 October, just before dawn, 150 Chinese guns opened an artillery barrage on the Thag-la Ridge. Ranged against the 20,000 Chinese assault troops were a mere 600 Indian defenders. Dalvi and the 7th Infantry Brigade were swept aside, those not killed being taken into captivity in China. In panic, the Indian government now called on Britain and the USA for military aid. But with Chinese troops poised to invade Assam and sweep down on to the plains of India, their leaders in Beijing abruptly called a halt to hostilities and agreed to withdraw from the Northeast Frontier Agency. They had made their point – further fighting was unnecessary.

OPERATION 'EAGLE CLAW' (1980)

Is it possible for a military operation to be too carefully planned? One would think not. Yet in an age when soldiers – like people in almost every other walk of life – have come to lean heavily on technology, it is possible that some of the simpler military virtues may be overlooked. Flexibility – the knack of making the most of what occurs naturally, rather than trying to impose your own order on what is essentially disordered – has always been a quality close to the heart of history's great generals. Plans that are too rigid run the risk of being destroyed by chance events. This was the real reason for the failure of the American attempt to rescue the hostages held in Iran in 1980.

On 4 November 1979, the American embassy in Teheran was attacked and overrun by militant followers of the Ayatollah Khomeini. They took prisoner a total of 66 Americans and held them hostage in a bid to force the USA to hand over the deposed Shah, then in exile in Egypt. Emotions ran high in both America and Iran, yet throughout the crisis, President Jimmy Carter publicly rejected the use of force and insisted on looking for a diplomatic solution. In fact, Carter was lying. Straight after the seizure of the embassy he ordered plans to be made for a rescue effort. Yet was a 'rescue' possible? Any kind of military action against Iran would carry enormous risks. The reaction of the Soviet Union, having only recently invaded Afghanistan, could not be predicted. Nor could the lives of the hostages be guaranteed in the event of an open war against Iran. For Carter only a 'surgical' operation, cleanly carried out, with a minimum of damage and casualties could possibly be justified. Yet he and his advisers were fooling themselves. The kind of operation they wanted was impossible from the start.

Operation Eagle Claw was just about the most difficult military operation of its size ever mounted. In fact, so difficult was the task it set itself, that it seems to belong more to the world of Hollywood, where script writers can control the numerous external factors that might otherwise ruffle the hair of the leading man.

President Carter and his top brass were asking American combat ships, planes and troops to assemble in force, and yet unnoticed, in the Persian Gulf, one of the busiest waterways in the world. Then they were to infiltrate assault troops into Iran secretly. Having achieved this, they were to fly about a thousand miles through the airspace of a state hostile to the USA, before landing, again unnoticed, near a city of over three million people. The soldiers were then to locate the American hostages in downtown Teheran, free them from the clutches of their fanatical captors, escape through the streets of a hostile city, crowded with people baying for the blood of Uncle Sam, arrange pick-ups by waving to waiting American helicopters and then fly at least four hundred miles to safety, presumably again unnoticed by passing squadrons of Iranian fighter planes or by ground-based missile operators. To make their task more interesting, the American troops were to achieve all this without inflicting casualties on the Iranian people with whom, as President Carter explained, the USA had absolutely no quarrel. It would have made a very good film but was poor stuff for the real world of the 1980s.

Led by a Rambo or an Indiana Jones it might have stood a good chance, but commanded by mere mortals it was in the snowflake-in-hell league. For a start Colonel 'Charging Charlie' Beckwith had been equipped with an 'abort button' which allowed him to cancel the operation if the plan began to go wrong. In such an improbable enterprise chance was bound to play a part and things would certainly go wrong. The plan should have been flexible enough to allow for the unexpected. Under the enormous stress of the operation a commander was always going to be too close to events to make the difficult decision on whether to stop or go on.

Beckwith's troops set off in six C-130 transport planes from an island off Oman to rendezvous in Iran at Desert Base One with eight Sea Stallion

The grim aftermath of an impossible mission – wreckage of a burned-out American helicopter litters the eastern Iranian desert. The Carter administration's plan to rescue 66 American hostages seized by Khomeiniite students in Teheran in 1980 placed great emphasis on technology at the expense of flexibility, and the mission had to be aborted because of equipment failure.

helicopters from the aircraft carrier *Nimitz*. The helicopters would lift the assault troops on to a second desert base and then to a site on the outskirts of Teheran, from which the 'snatch' would be made. From the start the helicopters encountered every kind of difficulty. They were committed to absolute radio silence and were therefore unable to communicate with each other when things went wrong. Although flying together in a diamond formation, every chopper was really on its own.

Crossing into Iranian territory, Chopper No. 8 began to develop trouble in its rotor gearbox. Radio silence meant its pilot could not tell anyone. He just gritted his teeth and flew on. Hugging the ground to avoid Iranian radar, the eight choppers were in constant danger of hitting the craggy terrain. Next Chopper No. 6 was forced to land with a damaged rotor blade. Luckily the pilot of Chopper No. 8 saw this happen and was able to land alongside, picking up No. 6's crew, and setting off in the wake of the remaining six helicopters, now some miles ahead.

Things became really difficult when the helicopters encountered a 'haboob', a cloud of sand and dust, thousands of feet high and miles in width.

The pilots had no option but to keep flying blindly through this fog, hoping not to hit anything. The Americans, in fact, had overestimated the efficiency of Iranian radar and were risking disaster by asking their choppers to fly at just 100 feet, since the fog was thickest near to the ground. To the pilots it was like 'flying in a bowl of milk', and the dust soon got inside the cockpits, raising temperatures, coating goggles and making breathing difficult. Having succeeded in passing through the first 'haboob' they then had the misfortune to encounter a second, even bigger one. There had been no mention of this kind of problem in their briefing. There was no other option but to plunge on into the fog again. Meanwhile, Chopper No. 5 was in trouble, with its navigation instruments malfunctioning. It had no option but to turn around and fly back to the *Nimitz*. The Americans were now down to just six helicopters and the mission plan had said that this was the minimum necessary for the operation to continue. Any further loss and it would be up to Beckwith to cancel the whole show.

Meanwhile, Colonel Beckwith and his men had arrived at Desert Base One and had immediately encountered two unexpected problems. The C-

130s had just landed near a main road when a bus full of Iranians came into sight. When challenged the bus refused to stop and the Americans were forced to shoot out its tyres and radiator. The passengers were ordered to leave the bus at gunpoint and a crowd of 44 Iranians, many of them women and children, climbed out. They were held some distance away and become eyewitnesses to the extraordinary and tragic events that followed. At this stage Beckwith was unflurried, quipping, 'I wasn't going to worry until we stopped ten buses. Then we'd have a parking problem.' However, his smile must have dimmed somewhat when a fuel lorry thundered into sight and crashed through the American checkpoints. The Americans fired an anti-tank weapon at the vehicle and the night was illuminated by an explosion as the lorry burst into flames. The driver, a tiny Iranian with nine lives, somehow managed to leap out and was picked up by a partner travelling behind in a small truck. To the astonishment of the Americans, the pair veered off the road and set off across open country. (The men were in fact smugglers, who took the Americans for Iranian police and were unlikely to report what had just happened.)

No amount of training could have prepared the Americans for the problems that now arose. After landing the engines of the planes and of the helicopters had been kept running to prevent contamination and malfunction in the sandy conditions. But the noise made communication either by word of mouth or radio almost impossible. In the darkness, with identification difficult, it became a problem to locate the commanders. One soldier apparently stood talking to a group of men whom he presumed were from Chopper No. 6 before realizing that they were Iranian bus passengers.

The arrival shortly afterwards of the six remaining helicopters had a dampening effect on Beckwith's morale. The pilots were clearly exhausted and in no condition to continue the mission. When he heard that Chopper No. 2 had hydraulic problems he decided that the show was

off. At this stage General Vaught, the Joint Task Force Commander, was contacted to discuss the emergency. Although leaving the final decision to the man on the spot, Vaught asked Beckwith to go ahead with just five helicopters, but the colonel refused. There was nothing for it but to get everyone out as quickly as possible.

As the Americans began to refuel the helicopters for withdrawal disaster struck. One of the choppers struck a tanker plane and its rotor blades sliced off the cockpit canopy. There was a huge explosion and both aircraft burst into flames. At once ammunition and rockets began to ignite, firing in all directions. Many Americans believed they were under attack. Eight men died in the terrible inferno that had engulfed the C-130 and the helicopter. Everywhere was confusion. Soldiers abandoned their equipment in a mad stampede to escape. All the helicopters were abandoned, complete with their secret maps and equipment. The desert was strewn with all the signs of a defeated army. As the puzzled Iranian bus passengers looked on, the surviving C-130s lifted off five wounded men and the rest of Beckwith's assault group.

The following morning President Carter announced the fiasco to an astounded world. The Iranians responded by bombing and strafing the site of Desert Base One, before landing and picking through the debris like a convention of scrap metal dealers. American television audiences had to endure the stomach-churning antics of Ayatollah Khalkali, who, with a disregard for normal human values, picked up the charred bones of the dead American airmen and held them up to the cameras.

Could Operation Eagle Claw have worked? We will never know. But so unlikely was a successful outcome that it was probably best that it ended at Desert Base One. In crowded Teheran the chances of failure would have been so high that it is easy to agree with one of the hostages who, on hearing of what happened, commented, 'Thank God for the sandstorm.'

INDEX

PICTURE ACKNOWLEDGEMENTS

The publishers wish to thank the following for permission to reproduce pictures in this book:

Archiv für Kunst und Geschichte
Associated Press Ltd
Bettman Archive
The Hulton Picture Company
Imperial War Museum
Mary Evans Picture Library
National Army Museum
Popperfoto